A Primer of
Abstract Mathematics

A Primer of
Abstract Mathematics

Robert B. Ash

Published by

THE MATHEMATICAL ASSOCIATION OF AMERICA

CLASSROOM RESOURCE MATERIALS

Classroom Resource Materials is intended to provide supplementary classroom material for students—laboratory exercises, projects, historical information, textbooks with unusual approaches for presenting mathematical ideas, career information, etc.

101 Careers in Mathematics, edited by Andrew Sterrett
Calculus Mysteries and Thrillers, R. Grant Woods
Combinatorics: A Problem Oriented Approach, Daniel A. Marcus
Elementary Mathematical Models, Dan Kalman
Interdisciplinary Lively Application Projects, edited by Chris Arney
Laboratory Experiences in Group Theory, Ellen Maycock Parker
Learn from the Masters, Frank Swetz, John Fauvel, Otto Bekken, Bengt Johansson, and Victor Katz
Mathematical Modeling for the Environment, Charles Hadlock
A Primer of Abstract Mathematics, Robert B. Ash
Proofs Without Words, Roger B. Nelsen
A Radical Approach to Real Analysis, David M. Bressoud
She Does Math!, edited by Marla Parker

MAA Service Center
P. O. Box 91112
Washington, DC 20090-1112
1-800-331-1622   fax: 1-301-206-9789

Preface

The purpose of this book is to prepare you to cope with abstract mathematics. The intended audience consists of: prospective math majors; those taking or intending to take a first course in abstract algebra who feel the need to strengthen their background; and graduate students (and possibly some undergraduates) in applied fields who need some experience in dealing with abstract mathematical ideas. If you have studied calculus, you have had some practice working with common functions and doing computations. If you have taken further courses with an applied flavor, such as differential equations and matrix algebra, you have probably begun to appreciate mathematical structure and reasoning. If you have taken a course in discrete mathematics, you may have some experience in writing proofs. How much of this is sufficient background for the present text? I don't know; it will depend on the individual student. My suggestion would be that if you have taken some math courses, enjoyed them and done well, give it a try.

Upon completing the book, you should be ready to handle a first course in abstract algebra. (It is also useful to prepare for a first course in abstract analysis, and one possible source is *Real Variables With Basic Metric Space Topology* by Robert B. Ash, IEEE Press, 1993. This basic analysis text covers the course itself as well as the preparation.)

In studying any area of mathematics, there are, in my view, three essential factors, in order of importance:

1. Learning to think intuitively about the subject;
2. Expressing ideas clearly and cogently using ordinary English;
3. Writing formal proofs.

Abstract language is used by mathematicians for precision and economy in statements and proofs, so it is certainly involved in item 3 above. But abstraction can interfere with the learning process, at all levels, so for best results in items 1 and 2, we should use abstract language sparingly. We are pulled in opposite directions and must compromise. I will try to be as informal as I can, but at some point we must confront the beast (i.e., an abstract theorem and its proof). I think you'll find that if you understand the intuition behind a mathematical statement or argument, you will have a much easier time finding your way through it.

I've attempted to come up with a selection of topics that will help make you very comfortable when you begin to study abstract algebra. Here is a summary:

1. Logic and Foundations.   Basic logic and standard methods of proof; sets, functions and relations, especially partial orderings and equivalence relations.

2. Counting.   Finite sets and standard methods of counting (permutations and combinations); countable and uncountable sets; proof that the rational numbers are countable but the real numbers are uncountable.

3. Elementary Number Theory.   Some basic properties of the integers, including the Euclidean algorithm, congruence modulo m, simple diophantine equations, the Euler φ function, and the Möbius Inversion Formula.

4. Some Highly Informal Set Theory.   Cardinal numbers and their arithmetic; well-ordering and its applications, including Zorn's Lemma.

5. Linear Algebra.   Finite-dimensional vector spaces, along with linear transformations and their representation by matrices.

6. Theory of Linear Operators.   Jordan Canonical Form; minimal and characteristic polynomials; adjoints; normal operators.

A single chapter on a subject such as number theory does not replace a full course, and if you find a particular subject interesting, I would urge you to pursue the area further. The more mathematics you study, the more skillful you will become at it.

Another purpose of the book is to provide one possible model for how to write mathematics for an audience with limited experience in formalism and abstraction. I try to keep proofs short and as informal as possible, and to use concrete examples which illustrate all the features of the general case. When a formal development would take too long (notably in set theory), I try to replace the sequence of abstract definitions and theorems by a consistent thought process. This makes it possible to give an intuitive development of some major results. In the last chapter on linear operators, you are given a powerful engine, the Jordan Canonical Form. The proof of existence is difficult and should probably be skipped on first reading. But using the Jordan form right from the start simplifies the development considerably, and this should contribute to your understanding of linear algebra.

Each section has a moderate number of exercises, with solutions given at the end of the book. Doing most of them will help you master the material, without (I hope) consuming too much time.

The book may be used as a text for a course in learning how to think mathematically. The duration of the course (one semester, one quarter, two quarters) will depend on the background of the students. Chapter 3, Chapter 4, and Chapters 5–6 are almost independent. (Before studying Chapter 5, it is probably useful to look at the description of various algebraic structures at the beginning of Section 3.3 and the definition of a vector space at the end of Section 4.2.) A shorter course can be constructed by choosing one or two of these options after covering Chapters 1 and 2.

We are doing theoretical, abstract mathematics, and students in applied fields may wonder where the applications are. But a computer scientist needs to know some elementary number theory in order to understand public key cryptography. An electrical engineer might want to study basic set theory in order to cope with abstract algebra and thereby learn about error-correcting codes. A statistician needs to know some theoretical linear algebra (projections, diagonalization of symmetric matrices, quadratic forms) in order to work with the multivariate normal distribution. There is potentially a large audience for abstract mathematics, and to reach this audience it is not necessary for us to teach detailed physical and engineering applications. The physics and engineering departments are quite capable of doing this. It is certainly useful to suggest possible applications, and as an illustration, I have included an appendix giving a typical application of linear algebra. But it is essential that we write in an accessible and congenial style, and give informal or heuristic arguments when appropriate.

Some acknowledgments: I got the idea of doing an intuitive development of set theory after seeing an informal discussion of the Maximum Principle in *Topology, A First Course* by James R. Munkres, Prentice-Hall 1975. I thank Ed Merkes for many helpful suggestions to improve the exposition, Ken Ross and Andy Sterrett for their encouragement and advice, and my wife Carol Ash for many insights on the teaching of combinatorics and linear algebra.

A typical reader of this text is likely to be motivated by a need to deal with formal mathematics in his or her professional career. But I hope that in addition there will be some readers who will simply take pleasure in a mathematical journey toward a high level of sophistication. There are many who would enjoy this trip, just as there are many who might enjoy listening to a symphony with a clear melodic line.

Robert B. Ash

Contents

Chapter 1 Logic And Foundations 1

    1.1   Truth Tables 1
    1.2   Quantifiers 6
    1.3   Proofs. 8
    1.4   Sets. 11
    1.5   Functions 14
    1.6   Relations 18

Chapter 2 Counting 25

    2.1   Fundamentals. 25
    2.2   The Binomial and Multinomial Theorems. 32
    2.3   The Principle of Inclusion and Exclusion. 34
    2.4   Counting Infinite Sets. 40

Chapter 3 Elementary Number Theory 45

    3.1   The Euclidean Algorithm. 45
    3.2   Unique Factorization. 48
    3.3   Algebraic Structures. 51
    3.4   Further Properties of Congruence Modulo m. 55
    3.5   Linear Diophantine Equations and Simultaneous Congruences. 57
    3.6   Theorems of Euler and Fermat. 61
    3.7   The Möbius Inversion Formula. 63

Chapter 4 Some Highly Informal Set Theory 69

    4.1   Well-Orderings. 69
    4.2   Zorn's Lemma and the Axiom of Choice. 72
    4.3   Cardinal Numbers. 74
    4.4   Addition and Multiplication of Cardinals. 78

Chapter 5   Linear Algebra 81

5.1   Matrices......81
5.2   Determinants and Inverses......86
5.3   The Vector Space F^n; Linear Independence and Bases......92
5.4   Subspaces......96
5.5   Linear Transformations......102
5.6   Inner Product Spaces......108
5.7   Eigenvalues and Eigenvectors......114

Chapter 6   Theory of Linear Operators 123

6.1   Jordan Canonical Form......123
6.2   The Minimal and Characteristic Polynomials......127
6.3   The Adjoint of a Linear Operator......131
6.4   Normal Operators......134
6.5   The Existence of the Jordan Canonical Form......141

Appendix: An Application of Linear Algebra 145

Solutions to Problems 147

List of Symbols 177

Index 179

1

Logic and Foundations

1.1 Truth Tables

Mathematicians must frequently deal with assertions known as *propositions;* these are statements that are either true or false. Sometimes it is very easy to determine truth or falsity (for example, $2 < 5$ is true), and sometimes it may be very complicated. But if we have some propositions whose truth value is known, there is a mechanical procedure to determine the truth value of new propositions that are formed from the old using the connectives "or", "and", and "not". The method uses *truth tables,* which we now describe.

Suppose A and B are propositions, and we form the proposition "A or B", which is regarded as true when *either A or B* is true. (In mathematics, "A or B" always means "A or B or both", unless you specifically say otherwise.) We then have the following truth table:

A	B	A or B
T	T	T
T	F	T
F	T	T
F	F	F

Thus the only time "A or B" will be false occurs when A and B are both false.

Suppose we have three propositions A, B, C; what does "A or B or C" mean? The "or both" idea is a bit awkward, but let's rephrase the definition of A or B. The proposition "A or B" is true when at least one of the assertions A, B is true. This generalizes in a very natural way to an arbitrary number of propositions. In particular, "A or B or C" is regarded as true when *at least one* of A, B, C is true. Here is the truth table:

A	B	C	A or B or C
T	T	T	T
T	T	F	T
T	F	T	T
T	F	F	T
F	T	T	T
F	T	F	T
F	F	T	T
F	F	F	F

Again, the only time the "or" proposition is false occurs when all the component propositions are false.

Now let's look at the "and" connective. The proposition "A and B" is defined to be true when *both* A and B are true. If we have three propositions A, B, C, then the assertion "A and B and C" is true when *all* the component statements are true. Here are the truth tables:

A	B	A and B
T	T	T
T	F	F
F	T	F
F	F	F

A	B	C	A and B and C
T	T	T	T
T	T	F	F
T	F	T	F
T	F	F	F
F	T	T	F
F	T	F	F
F	F	T	F
F	F	F	F

The "not" connective causes no difficulty: "not A" is true when A is false, and false when A is true. The truth table has only two lines:

A	not A
T	F
F	T

Mathematicians make a living by proving theorems. A theorem is an assertion that a conclusion B follows from a collection A of hypotheses, in other words, a statement of the form

$$A \text{ implies } B.$$

Another way to say this is to claim that if A holds, then it is guaranteed that B will hold as well, i.e.,

$$\text{If } A \text{ then } B.$$

Yet another version is to state that the hypotheses A are sufficient to assure that B occurs, i.e.,

$$A \text{ is } sufficient \text{ for } B.$$

Still another way: if A holds, then it is necessarily the case that B follows, i.e.,

$$B \text{ is } necessary \text{ for } A.$$

Finally, we can say that A holds only if B is also satisfied; for short,

$$A \text{ only if } B.$$

Now "A implies B" is a compound statement that is constructed from the components A and B, and we should be able to make a sensible definition of its truth value if we know the truth values of A and B. How do we capture the idea of implication? One idea is to ask the question "How can the statement 'If A then B' fail?" For B not to follow from A, there must be a case where A is true but B is false. This leads to the following truth table:

A	B	A implies B
T	T	T
T	F	F
F	T	T
F	F	T

Notice that we have taken "A implies B" to be true if A is false, regardless of the truth value of B. In this case, we have not produced a counterexample, in other words a situation in which A holds but B doesn't. There is another argument that you might find more convincing. Consider "If A then B" where A is "n is divisible by 6" and B is "n is divisible by 3"; here, n is a fixed positive integer. It seems sensible to hope that in any reasonable mathematical system, "A implies B" should hold for all n. But suppose $n = 9$; then A is false but B is true; and if $n = 8$, then both A and B are false. If you want "A implies B" to be true regardless of the particular value of n, you are forced to enter T in the last two lines of the right-hand column of the above truth table.

There is some standard terminology that is conveniently discussed at this point. The *converse* of the assertion "A implies B" is "B implies A", and as above, we have several ways of expressing this:

If B then A;

A is *necessary* for *B;*

A if B.

Notice that if an assertion is true, it does not follow that the converse is true. If you live in the United States, then you live in North America, but not conversely. But if "A implies B" and "B implies A" are both true, then we say that A and B are equivalent propositions. In other words,

A is *equivalent* to *B*

means

$(A$ implies $B)$ and $(B$ implies $A)$.

Another way to say this is

A if and only if B.

Yet another way is

A is *necessary and sufficient* for *B.*

Now if A and B are indeed equivalent propositions, they should have the same truth value. In other words, the truth table for "A is equivalent to B" should be:

A	B	A is equivalent to B
T	T	T
T	F	F
F	T	F
F	F	T

We can verify this as follows. If both A and B are true or both are false, then both A implies B and B implies A are true, so [(A implies B) and (B implies A)] is true. On the other hand, if A is true and B is false, then A implies B is false, and if A is false and B is true, then B implies A is false. In either case, [(A implies B) and (B implies A)] is false.

The phrase "if and only if" is often abbreviated as *iff*. When making definitions, it is standard practice to say something like "A set S is said to have property P if condition C holds". The "if" in this case stands for "iff"; in other words, S has property P if and only if condition C holds.

Other standard abbreviations are

$$A \vee B \quad \text{for} \quad A \text{ or } B;$$

$$A \wedge B \quad \text{for} \quad A \text{ and } B;$$

$$A \Rightarrow B \quad \text{for} \quad A \text{ implies } B;$$

$$A \Leftrightarrow B \quad \text{for} \quad A \text{ is equivalent to } B;$$

$$\neg A \quad \text{for} \quad \text{not } A.$$

The notion of equivalence can be extended to *compound propositions,* which are constructed from individual propositions A, B, C,... by using the connectives "or", "and" and "not". For example, we have the *DeMorgan Laws*

(D1) $$\left[\neg(A \wedge B) \right] \Leftrightarrow \left[(\neg A) \vee (\neg B) \right]$$

(D2) $$\left[\neg(A \vee B) \right] \Leftrightarrow \left[(\neg A) \wedge (\neg B) \right].$$

What this means is that regardless of the truth or falsity of A and B, the left and right sides have exactly the same truth value. Thus, in the case of (D1), $\neg(A \wedge B)$ has exactly the same truth table as $(\neg A) \vee (\neg B)$, so that $\neg(A \wedge B)$ and $(\neg A) \vee (\neg B)$ are equivalent propositions. Let's check this:

A	B	$\neg A$	$\neg B$	$\left[(\neg A) \vee (\neg B) \right]$	$A \wedge B$	$\neg(A \wedge B)$
T	T	F	F	F	T	F
T	F	F	T	T	F	T
F	T	T	F	T	F	T
F	F	T	T	T	F	T

which is as expected. It is also profitable to reason directly. For not(A and B) to be true, (A and B) must be false, which means that either A is false or B is false, i.e., (not A) or (not B) is true. A similar analysis can be carried out for (D2): not(A or B) will be true if and only if (A or B) is false, which happens if and only if both A and B are false, i.e., (not A) and (not B) is true.

The purpose of using parentheses and brackets in the above formulas is to avoid ambiguity. For example, $\neg A \vee B$ might mean (not A) or B, or perhaps not(A or B). But if we write $(\neg A) \vee B$, there is no doubt that we perform the "not" operation first, and then the "or". It is possible to give rules of priority so that parentheses become unnecessary, but we will not be using stylized formal language often enough to bother with this.

When I was a student in college, I took a course in symbolic logic. The textbook described a courtroom scene in which the defense attorney was presenting the final summation, which went as follows:

> Today is Thursday
> Today is not Thursday
> Therefore my client is innocent!

The judge peered down from the bench and said "With all due respect, counselor, that is not much of an argument". The defense counsel replied "On the contrary, the argument is perfectly valid. It is of the form 'A implies B' where A (Today is Thursday and today is not Thursday) is known to be false." The vignette ended at this point, leaving the impression that logic had little to do with real-life experience. It was not until years later that I realized that both the author of the text and the instructor had missed the main point. Although admittedly, A implies B is true, this reveals nothing whatever about the truth or falsity of B, which is the issue before the court.

An implication that is guaranteed to be true because the hypothesis is always false is sometimes said to be *vacuously true*.

Finally, a *tautology* is a proposition that is always true, regardless of the truth values of its components. A *contradiction* is a proposition that is always false.

Problems For Section 1.1

1. Prove the second DeMorgan Law, i.e., $\neg(A \vee B)$ is equivalent to $(\neg A) \wedge (\neg B)$.

2. Extend the DeMorgan Laws to n propositions, as follows:

$$\left[\neg(A_1 \wedge A_2 \wedge \cdots \wedge A_n)\right] \Leftrightarrow \left[(\neg A_1) \vee (\neg A_2) \vee \cdots \vee (\neg A_n)\right]$$
$$\left[\neg(A_1 \vee A_2 \vee \cdots \vee A_n)\right] \Leftrightarrow \left[(\neg A_1) \wedge (\neg A_2) \vee \cdots \vee (\neg A_n)\right].$$

3. Show that "A implies B" is equivalent to "(not A) or B".

4. Show that "A or (not A)" is a tautology, and that "A and (not A)" is a contradiction.

5. Show that $\left[A \wedge (B \vee C)\right] \Leftrightarrow \left[(A \wedge B) \vee (A \wedge C)\right]$.

6. Show that $\left[A \vee (B \wedge C)\right] \Leftrightarrow \left[(A \vee B) \wedge (A \vee C)\right]$.

7. Suppose that the compound propositions P and Q are formed using only $\vee$, $\wedge$ and $\neg$. (The implication $A \Rightarrow B$ can be replaced by $(\neg A) \vee B$; see Problem 3.) Suppose that P and Q are known to be equivalent. Now interchange $\vee$ and $\wedge$ to obtain propositions P^* and Q^*. (If a tautology T or a contradiction F appears in the original propositions, interchange T and F.) The new propositions P^* and Q^* are still equivalent; can you see why?

(Suggestion: Look at Problems 5 and 6 for a specific example. Apply the DeMorgan Laws to P and Q, and the result will look almost like P^* and Q^*, except that instead of A, we have $\neg A$, and similarly for B and C. But A is an *arbitrary* proposition.)

Mathematicians express the idea that the "or" and "and" connectives can be interchanged without affecting truth by referring to the *duality* between the two connectives. A similar duality occurs between union and intersection of sets, to be examined in Section 1.4.

1.2. Quantifiers

Not all mathematical statements are propositions which have a definite truth value. For example, consider the statement

$$x \text{ is less than } 5$$

where x is a real number. The assertion is true for some values of x and false for others. It is possible to convert such a statement into a proposition by using quantifiers, which are of two types, existential and universal.

The *existential quantifier* $\exists$ means "there exists"; other equivalent phrases are "there is at least one" and "for some".

The *universal quantifier* $\forall$ means "for all", equivalently, "for every".

Thus, again thinking of x as a real number,

$$\exists x \ (x < 5)$$

means that there exists an x such that $x < 5$; in other words, there is at least one real number that is less than 5. This is certainly true. On the other hand,

$$\forall x \ (x < 5)$$

means that for all x, $x < 5$; in other words, every real number is less than 5. This is definitely false.

More than one quantifier can appear in a sentence. For example, with x and y real,

$$\forall x \ \exists y \ (x + y = 13) \tag{1}$$

says that for every x there is a y such that $x + y = 13$. This is true; if you give me x, I can solve for y: $y = 13 - x$. However, consider the assertion

$$\exists x \ \forall y \ (x + y = 13). \tag{2}$$

This says that for some x, *every* y satisfies $x + y = 13$; definitely false.

There is a very convenient mechanical procedure for finding the negation of a statement involving quantifiers. Here is how it works in a typical case; let's analyze the assertion

$$\text{not} \ \big[(\exists x) \ (x < 5)\big].$$

We are saying that it is *not* the case that some x is less than 5; in other words, *every x* is greater than or equal to 5. Thus we have

$$\forall x \ (x \geq 5).$$

Thus to go from $\exists x \ (x < 5)$ to its negation, we have reversed the quantifier (from $\exists$ to $\forall$) and changed the main statement $x < 5$ to its negative. Similarly, to negate $\forall x \ (x \geq 5)$, i.e., to say that "every x is at least 5" is false; we assert that some x is less than 5. So we arrive at

$$\exists x \ (x < 5),$$

which is not a surprise. Taking the negation twice brings us right back to the original statement.

This process works when there is more than one quantifier. For example, to negate (1) above, notice that it is of the form $\forall x \ P$, so that its negative is $\exists x \ (\text{not } P)$. But P is $\exists y \ (x + y = 13)$, so not P is $\forall y \ (x + y \neq 13)$. Thus the negation of (1) is

$$\exists x \ \forall y \ (x + y \neq 13),$$

which illustrates how the method works in general. Proceed from left to right, reversing quantifiers as you go, until you reach the main statement, which you change to its negation.

The truth or falsity of a statement involving quantifiers depends crucially on the allowable values of the variables. For example, suppose we change (1) slightly:

$$\forall x \ (\exists y > 0) \ (x + y = 13) \tag{3}$$

which says that for every real x there is a *positive* real y such that $x + y = 13$. This is false, e.g., take x to be 13 or larger. Taking the negation of (3) gives

$$\exists x \ (\forall y > 0) \ (x + y \neq 13), \tag{4}$$

which says that there is at least one real x such that every positive y satisfies $x + y \neq 13$. This is true; again, take x to be 13 or larger. If you add a positive number to x, you certainly can't get 13.

Caution. The phrase "> 0" in (3) and (4) refers to the allowable values of the variable y. It is *not changed* when we go from a statement to its negation.

Perhaps you recall the definition of a limit from calculus, and perhaps you remember being baffled by it. The limit statement is actually a very complicated mathematical sentence using three quantifiers, and it is not surprising that it causes confusion. What does it mean to say the sequence $x_1, x_2, \ldots$ of real numbers converges to the real number x? Intuitively, as n gets large, x_n gets very close to x. How close? As close as you wish. For example, suppose we want the sequence to be within 10^{-9} of x, i.e., $x - 10^{-9} < x_n < x + 10^{-9}$. How large must n be? It might turn out in a particular case that all x_n's from $n = 10^{15}$ onward satisfy this inequality. Thus for *every* degree of closeness you might give me, there is *some* point in the sequence so that *every* x_n from that point on is that close to x. The degree of closeness is measured by a small positive number ε, 10^{-9} in this case. The point in the sequence that achieves the desired closeness is measured by a positive integer N; in this case, $N = 10^{15}$. We can then write the formal definition of convergence of x_n to x:

$$x_n \to x \text{ means } (\forall \epsilon > 0)(\exists N)(\forall n \geq N)(|x_n - x| < \epsilon).$$

For every positive real number ϵ there is a positive integer N such that for every positive integer $n = N, N + 1, N + 2, \ldots, x_n$ differs from x by less than ϵ.

If you did not understand the definition of a limit in calculus and it makes a bit more sense now, fine. If it is still obscure, there is no problem; we will not be using it. I brought it up to make the point that quantifiers will be encountered in all areas of mathematics.

Problems For Section 1.2

1. Express the statement "For every real number there is a larger integer" using quantifiers.

2. Take the negation of the statement in Problem 1, using the mechanical procedure given in the text. Express the statement in ordinary English, and verify that it is false.

1.3. Proofs

The technique used to prove a particular theorem will usually depend on the specific mathematical objects being considered. Proofs in algebra will have a different flavor from those in probability or complex variables. But there are certain basic patterns that occur over and over again, regardless of the area of study. So it's useful to make a list of these basic patterns and to be aware of them when planning to prove a result.

Suppose that we are trying to prove that A implies B; here are some ideas.

1. The Direct Proof. This can be described very quickly; we assume the hypothesis A, and then deduce the conclusion B. Easier said than done, of course, but the point is that we work directly with the propositions A and B rather than with compound propositions formed from them, as in the techniques to follow.

2. Proof of the Contrapositive. The *contrapositive* of the proposition A implies B is

$$(\text{not } B) \text{ implies } (\text{not } A),$$

and any implication is equivalent to its contrapositive. In more formal language,

$$[(A \Rightarrow B)] \Leftrightarrow [(\neg B) \Rightarrow (\neg A)].$$

As we know from Section 1.1, this means that $(A \Rightarrow B)$ and $[(\neg B) \Rightarrow (\neg A)]$ have exactly the same truth value. Rather than construct the entire truth table, let's take a short cut. The only time $A \Rightarrow B$ will be false occurs when A is true and B is false; the only time $(\neg B) \Rightarrow (\neg A)$ will be false occurs when $(\neg B)$ is true and $(\neg A)$ is false, that is, B is false and A is true. Therefore the implication and its contrapositive are equivalent, so that to prove that A implies B, we are free to switch to the contrapositive. How do we know when to make the switch? If we are having trouble constructing a direct proof, we might try the contrapositive, hoping to make things easier.

3. Proof by Contradiction. Under a hypothesis A, we are trying to show that a proposition B is true. Suppose that we assume that B is false, and somehow arrive at a contradiction, in other words, a statement that is pure nonsense. For example, "It is Thursday and it is not Thursday". We can then conclude that B is true. To spell this out in more precise language:

Recall that a *contradiction* is a proposition that is *always false*. For our purposes, the specific content of the proposition is not of interest. All that matters is that we have a statement whose truth value is always F, so let's abbreviate such a statement simply by F. We are saying that if we prove that (not B) implies F, we are entitled to conclude B; in other words,

$$\left[(\neg B) \Rightarrow F\right] \Rightarrow B. \tag{1}$$

If our method is to be valid, the implication (1) must always be true; let's verify this using a truth table:

B	$\neg B$	F	$(\neg B) \Rightarrow F$	$\left[(\neg B) \Rightarrow F\right] \Rightarrow B$
T	F	F	T	T
F	T	F	F	T

as expected. An assertion that is *always true* is a *tautology;* it is the opposite of a contradiction, and can be abbreviated simply by T. Thus statement (1) is a tautology; another simple example of a tautology is $A \Rightarrow A$, which must be true regardless of the truth or falsity of A.

4. Proof by Cases. Imagine that we are doing a proof about real numbers. At some point in the argument, we might be worried about whether a particular number x is less than 5, equal to 5, or greater than 5. It might be convenient to consider the three cases separately. If we can establish the desired conclusion in each case, we are entitled to claim that the conclusion is always valid, because a real number x is either less than 5 or equal to 5 or greater than 5. Here is a general description of this situation:

If we prove that A_1 implies B, A_2 implies B, and A_3 implies B, it follows that (A_1 or A_2 or A_3) implies B. Using formal language,

$$\left[(A_1 \Rightarrow B) \wedge (A_2 \Rightarrow B) \wedge (A_3 \Rightarrow B)\right] \Rightarrow \left[(A_1 \vee A_2 \vee A_3) \Rightarrow B\right] \tag{2}$$

A similar assertion can be made if we have n cases $A_1, \ldots, A_n$. To show that (2) is a tautology, ask how (2) can be false. The only way is for the left side to be true and the right side to be false. The right side will be false if and only if at least one A_i is true but B is false. But then one of the implications $A_i \Rightarrow B$ on the left is false, so the entire left side is false. This guarantees that the implication (2) is true.

5. Mathematical Induction. This is often a useful technique when proving results that are to hold for every positive integer n. For example, let's prove that the sum of the first

n odd integers is n^2, that is,

$$1 + 3 + 5 + \cdots + (2n - 1) = n^2. \tag{3}$$

The idea is to show that (3) holds for $n = 1$ (the *basis* step), and then, assuming that the result holds for a given n (the *induction hypothesis*), prove it for $n + 1$. Then (3) must be true for all n, because of the domino effect. We know that our result is valid for $n = 1$; if it holds for $n = 1$, then it is true for $n = 2$, and therefore the $n = 2$ case is disposed of. Again, since we have $n = 2$ we must have $n = 3$; since we have $n = 3$ we must have $n = 4$, and so on. Now when $n = 1$, (3) says that the sum of the first odd integer is 1, in other words, $1 = 1$, which is true. Now if (3) holds for n, let's compute the sum of the first $n + 1$ odd integers:

$$1 + 3 + 5 + \cdots + (2n - 1) + (2n + 1) = n^2 + (2n + 1) \quad \text{by the induction hypothesis}$$

$$= (n + 1)^2 \quad \text{by algebra}$$

and therefore (3) holds for $n + 1$.

As another example, let's prove that $5^n - 4n - 1$ is divisible by 16 for all integers $n = 1, 2, 3, \ldots$.

Basis step: When $n = 1$, we have $5^n - 4n - 1 = 5 - 4 - 1 = 0$, which is divisible by 16.

Induction step: Our *induction hypothesis* is that $5^n - 4n - 1$ is divisible by 16. We must prove that the result holds for $n + 1$; in other words, we must show that $5^{n+1} - 4(n + 1) - 1$ is divisible by 16. This follows by a bit of clever algebra:

$$5^{n+1} - 4(n + 1) - 1 = 5(5^n - 4n - 1) + 16n.$$

By the induction hypothesis, $5^n - 4n - 1$ is divisible by 16, and certainly $16n$ is divisible by 16. Thus the sum, namely $5^{n+1} - 4(n + 1) - 1$, is divisible by 16.

Proofs by mathematical induction often work very smoothly, as in the above examples, but this does not reflect the considerable effort that might be required to discover the result that is to be proved.

Problems for Section 1.3

Use mathematical induction in all problems.

1. Prove that the sum of the first n positive integers is $n(n + 1)/2$.

2. Prove that $2^{2n} - 1$ is divisible by 3 for every positive integer n.

3. Prove that $11^n - 4^n$ is divisible by 7 for $n = 1, 2, \ldots$.

4. Show that the sum of the first n squares is $n(n + 1)(2n + 1)/6$, that is,

$$1^2 + 2^2 + \cdots + n^2 = \frac{n(n + 1)(2n + 1)}{6} \qquad \text{for } n = 1, 2, \ldots .$$

5. A post office sells only 5-cent and 9-cent stamps. Show that any postage of 35 cents or more can be paid using only these stamps. (In the induction step, do a proof by cases.

In case 1, postage of n cents can be paid using only 5-cent stamps, and in case 2, postage of n cents requires at least one 9-cent stamp.)

1.4 Sets

We cannot get very far in mathematics without dealing with sets and operations on sets. A *set* is simply a collection of objects (often referred to as *elements* or *points* of the set); if the number of elements is small, it may be feasible to list them all. Thus we write

$$A = \{a, b, c, d\}$$

to indicate that the set A consists of the four elements a, b, c, and d.

A set can often be described by giving a property that is satisfied by the elements. Thus we write

$$B = \{x \in \mathbb{R} : 2 \leq x \leq 3\}, \quad \text{where } \mathbb{R} \text{ is the set of real numbers,}$$

to indicate that B consists of those real numbers (that is, those x belonging to $\mathbb{R}$) between 2 and 3.

Just as the connectives "or", "and" and "not" allowed us to form new propositions from old, the same connectives enable us to form new sets from old. If A and B are sets, we define the *union* of A and B as

$$A \cup B = \text{ the set of elements belonging to } \textit{either } A \text{ or } B;$$

the *intersection* of A and B is given by

$$A \cap B = \text{ the set of elements belonging to } \textit{both } A \text{ and } B;$$

the *complement* of A is

$$A^c = \text{ the set of elements that do } \textit{not} \text{ belong to } A.$$

(When considering complements of sets, we normally assume that all elements belong to a fixed set Ω, called the *universe*. Then

$$A^c = \{x \in \Omega : x \notin A\},$$

that is, A^c consists of those elements x in Ω such that x does not belong to A.)

Unions and intersections can be defined for any number of sets. The *union* of the sets $A_1, A_2, \ldots, A_n$ is given by

$$A_1 \cup A_2 \cup \cdots \cup A_n = \text{ the set of elements belonging to } \textit{at least one} \text{ of the } A_i;$$

the notation $\bigcup_{i=1}^{n} A_i$ is also used. The *intersection* of $A_1, A_2, \ldots, A_n$ is defined by

$$A_1 \cap A_2 \cap \cdots \cap A_n = \text{ the set of elements belonging to } \textit{all} \text{ the } A_i;$$

the notation $\bigcap_{i=1}^{n} A_i$ is also used. (Complements are only defined for a single set.)

Unions and intersections can be defined equally well for an infinite number of sets. For example, we might have a set A_i for each real number i. The *union* of the A_i is

$$\bigcup_i A_i = \text{ the set of elements belonging to } A_i \text{ for at least one } i,$$

and the *intersection* of the A_i is

$$\bigcap_i A_i = \text{ the set of elements belonging to } A_i \text{ for all } i.$$

Pictures called *Venn diagrams* in which a set is represented by the interior of a circle can aid in visualizing operations on sets. For example, the intersection of A and B is the shaded area in Figure 1.4.1.

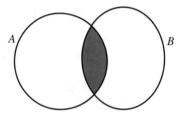

FIGURE 1.4.1
Intersection of A and B

Venn diagrams are useful when two or three sets are involved, but when the number of sets is four or more, they become too unwieldy. In abstract algebra, we often must deal with a very large finite number, or perhaps an infinite number, of sets, and for this reason, we will analyze statements about sets without using Venn diagrams. Instead, we will reason directly from the definitions of union, intersection, and complement.

In Section 1.1 we met the De Morgan Laws for propositions; there are analogous results for sets, with "or" corresponding to union, "and" to intersection, and "not" to complement.

De Morgan Laws For Sets.
  1. $\left(\bigcap_i A_i\right)^c = \bigcup_i (A_i^c)$
  2. $\left(\bigcup_i A_i\right)^c = \bigcap_i (A_i^c)$

The first law says the the complement of the intersection of arbitrarily many sets is the union of the complements. The second law states that the complement of the union is the intersection of the complements.

The De Morgan Laws follow quickly from the definitions of union, intersection and complement, and the technique of proof is essentially the same in both cases. Thus to prove the second law, observe that x belongs to $\left(\bigcup_i A_i\right)^c$ if and only if x does not belong to $\bigcup_i A_i$, that is, if and only if the statement "$x \in A_i$ for at least one i" is false. But this is equivalent to saying that for all i, we have $x \notin A_i$, which says that $x \in \bigcap_i (A_i^c)$. A similar analysis proves the first law (Problem 1).

Here is another example.

Distributive Law For Sets. $A \cap (\bigcup_i B_i) = \bigcup_i (A \cap B_i)$.

This resembles the distributive law of ordinary arithmetic if we allow intersection to correspond to multiplication and union to addition. Thus for numbers we have $a(b_1 + b_2 + \cdots) = ab_1 + ab_2 + \cdots$. But it would be unwise to push the analogy too far. For example, for sets we have $A \cup A = A \cap A = A$, but $a + a$ and aa do not equal a in general.

To prove the distributive law, suppose that x belongs to $A \cap (\bigcup_i B_i)$. Then $x \in A$ and $x \in B_i$ for at least one i; thus $x \in \bigcup_i (A \cap B_i)$. This argument may be reversed to show that if x belongs to $\bigcup_i (A \cap B_i)$, then $x \in A \cap (\bigcup_i B_i)$.

If we are trying to prove that two sets C and D are equal, and we show that if x belongs to C then x belongs to D, we are not finished yet. At this point we have shown that C is a *subset* of D; in other words, $x \in C$ implies $x \in D$. (We also say that C is *included* in D or *contained* in D.) It is conceivable that there are points in D that do not belong to C. To eliminate this possibility, we must prove that if $x \in D$, then $x \in C$.

The notation $C \subseteq D$ indicates that C is a subset of D. If C is strictly inside of D, that is, $C \subseteq D$ but $C \neq D$, we say that C is a *proper subset* of D, and write $C \subset D$.

It is useful to identify the smallest possible set, namely the set with no members. It is called the *empty set* and is denoted by $\varnothing$. The following result sometimes makes students a bit uncomfortable, but it is a consequence of our definitions.

For any set A, we have $\varnothing \subseteq A$.

To prove this, we must show that if $x \in \varnothing$, then $x \in A$. But this is vacuously true because "$x \in \varnothing$" is always false. To put it another way, if you find an element x in the empty set, I am confident that I will be able to prove that x belongs to A.

The sets A_i are said to be *pairwise disjoint* or *mutually exclusive* if no two sets have a point in common, in other words, $A_i \cap A_j = \varnothing$ for all $i \neq j$. We will usually omit the word "pairwise"; *disjoint* will be synonymous with *pairwise disjoint*.

If A and B are sets, the *difference* between A and B is defined by

$$A \setminus B = \{x : x \in A \text{ and } x \notin B\}.$$

That's enough set-theoretic terminology for now. Additional properties of sets will appear throughout the book, and we will deal with them as they arise.

Problems For Section 1.4

1.   Prove the first De Morgan Law for sets.

2.   The distributive law for sets is sometimes expressed by saying that intersection distributes over union. Show that equally well, union distributes over intersection; in other words, if we interchange union and intersection to obtain

$$A \cup \left(\bigcap_i B_i \right) = \bigcap_i (A \cup B_i),$$

the new identity is also valid.

3.   If $A \subseteq \varnothing$, show that $A = \varnothing$. (This may appear to be obvious, but prove it formally.)

4.   If the sets A_i are disjoint, and $B_i \subseteq A_i$ for all i, show that the sets B_i are disjoint.

5. If $A \cup B = A \cup C$, is it true in general that $B = C$?

6. Is it true that $A \cup (B \setminus A) = B$?

7. There is a valid identity of the form $A \cup (B \setminus A) = ?$ Replace ? by your candidate, and prove the result.

8. Suppose that we have a set identity of the form $C = D$, where C and D are formed from other sets using the operations of union, intersection and complement. Suppose that we interchange union and intersection to obtain sets C^* and D^*. (If the entire space Ω or the empty set $\varnothing$ appears in the original identity, interchange Ω and $\varnothing$.) Show that $C^* = D^*$. (Look at Problem 2 for a specific example.)

9. Show that $A \subseteq B$ if and only if $A \cap B = A$, and $A \subseteq B$ if and only if $A \cup B = B$. (A Venn diagram in which the circle representing A lies inside the circle representing B is convincing, but prove the results formally.)

1.5 Functions

We are on familiar ground here; you have worked with functions many times. For example, if $f(x) = x^2$, where x is a real number, then if I give you the value of x, you can square it to produce the value of $f(x)$, a nonnegative real number. In general, a *function* or *mapping* from a set A to a set B is a rule that assigns to each element x in A an element $f(x)$ in B. We write

$$f : A \to B;$$

A is called the *domain* of the function f, and B the *codomain*. In the above example, A is the set of reals, and B the set of nonnegative reals. Since a nonnegative real number is in particular a real number, we can if we like change B to the set of all reals. This does not change the fact that $f(x) = x^2$, but technically, B is part of the description of f, and a (fussy) mathematician would insist that we now have a (slightly) different function.

One of the most important operations on functions is that of composition, and again this is a familiar idea. For example, if $h(x) = \sin^2 x$, then given the value of x, we first compute $\sin x$ and then square the result. Thus if $f(x) = \sin x$ and $g(y) = y^2$, we have $h(x) = g(f(x))$. In general, if $f : A \to B$ and $g : B \to C$, the *composition* of f and g is the function $h : A \to C$ defined by $h(x) = g(f(x))$ for all $x \in A$. We write

$$h = g \circ f.$$

Figure 1.5.1 indicates that we compute h by first calculating f and then g.

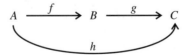

FIGURE 1.5.1
Composition of Functions

In mathematics we very often ask whether a particular function has a certain property. For example, if $f(x) = x^2$, x real, and I tell you the value of $f(x)$, does this determine x? Look at a concrete example; if $f(x) = 16$, then x can be 4, but it also can be -4. In general, if y is any positive real number, then there are two x's such that $x^2 = y$, namely $x = \sqrt{y}$ and $x = -\sqrt{y}$. If you draw the graph of f, the horizontal line at height y will intersect the graph in two points. On the other hand, if $y = 0$, there is only one value of x, namely $x = 0$; the graph touches the x-axis at only one point.

Here is another commonly asked question. If $f(x) = x^2$ and we take the codomain B to be the nonnegative reals, is B completely covered? In other words, if I give you any y in B, is there at least one x such that $f(x) = y$? The answer is yes, by the analysis that we just did. But suppose we had taken the codomain to be the entire set of reals. Now the answer is no, because if y is a negative real number, we cannot possibly have $y = x^2$ for any x.

This discussion leads us to the following formal statement.

1.5.1 Definitions.   The function $f : A \rightarrow B$ is said to be *injective* or *one-to-one* if no two distinct x's in A yield the same value of $f(x)$. In other words, if $x_1 \neq x_2$, then $f(x_1) \neq f(x_2)$. Equivalently (take the contrapositive), if $f(x_1) = f(x_2)$, then $x_1 = x_2$.

The function $f : A \rightarrow B$ is said to be *surjective* or *onto* if given any y in B, there is at least one x in A such that $f(x) = y$.

The function $f : A \rightarrow B$ is said to be *bijective* or a *bijection* or *one-to-one onto* or *a one-to-one correspondence* if f is both injective and surjective.

If the statement that $f(x) = y$ is represented by an arrow from x to y, then if f is not injective, there will be more than one arrow landing at the same point y. If f is not surjective, there will be points of B that are never hit by an arrow. See Figure 1.5.2 for an example.

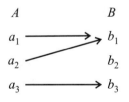

FIGURE 1.5.2
A function that is neither injective nor surjective

In this example we have $f(a_1) = f(a_2) = b_1$ and $f(a_3) = b_3$. The function f is not injective because $f(a_1) = f(a_2) = b_1$, and f is not surjective because there is no $a \in A$ such that $f(a) = b_2$.

Now if $f(x) = x^2$ from the reals to the nonnegative reals, then f is surjective but not injective. If $f(x) = x$ from the nonnegative reals to the reals, then f is injective but not surjective. But for *finite sets of the same size,* this situation cannot occur.

1.5.2 Theorem.   *If $f : A \to B$ where A and B each have n elements for some positive integer n, then f is injective if and only if f is surjective.*

Proof.   If f is not injective, there will be at least two arrows from distinct points of A to a single point of B. But then for f to be surjective, the remaining $n-1$ points of B must be covered by at most $n-2$ remaining arrows, which is impossible. We conclude that f cannot be surjective. Conversely, assume that f is not surjective, so that some point of B is not covered. Then we have n arrows from A to B landing on at most $n-1$ points, so there must be at least two arrows that land on the same point, so that f is not injective. ∎

If $f : A \to B$ and C is a subset of B, very often we will be interested in the set of points in A that map into C.

1.5.3 Definition.   Let $f : A \to B$, with $C \subseteq B$. The *preimage* of C under f is defined by

$$f^{-1}(C) = \{x \in A : f(x) \in C\}.$$

In Figure 1.5.2 we have

$$f^{-1}\{b_1\} = f^{-1}\{b_1, b_2\} = \{a_1, a_2\}, \quad f^{-1}\{b_2\} = \varnothing, \quad f^{-1}\{b_1, b_2, b_3\} = \{a_1, a_2, a_3\}.$$

In general, the preimage behaves extremely well with respect to the operations of union, intersection and complement.

1.5.4 Theorem.   *Let $f : A \to B$, and suppose we have an arbitrary collection of subsets B_i of B. Then*

$$f^{-1}\left(\bigcup_i B_i\right) = \bigcup_i f^{-1}(B_i);$$

$$f^{-1}\left(\bigcap_i B_i\right) = \bigcap_i f^{-1}(B_i);$$

$$f^{-1}(B_i^c) = \left[f^{-1}(B_i)\right]^c.$$

Thus the preimage of a union is the union of the preimages, the preimage of an intersection is the intersection of the preimages, and the preimage of a complement is the complement of the preimage.

Proof.   We have

$$x \in f^{-1}\left(\bigcup_i B_i\right) \quad \text{iff} \quad f(x) \in \bigcup_i B_i$$

$$\text{iff} \quad f(x) \in B_i \text{ for at least one } i$$

$$\text{iff} \quad x \in f^{-1}(B_i) \text{ for at least one } i$$

$$\text{iff} \quad x \in \bigcup_i f^{-1}(B_i);$$

the argument for intersections is exactly the same, with $\bigcup_i$ replaced by $\bigcap_i$ and "at least one" replaced by "all". Finally,

$$\begin{aligned}
x \in f^{-1}(B_i^c) \quad &\text{iff} \quad f(x) \in B_i^c \\
&\text{iff} \quad f(x) \notin B_i \\
&\text{iff} \quad x \notin f^{-1}(B_i) \\
&\text{iff} \quad x \in \left[f^{-1}(B_i) \right]^c. \ \blacksquare
\end{aligned}$$

If $f : A \to B$ and C is a subset of A, the *image* (or *direct image*) of C under f is defined as the set of all possible values $f(x)$ generated by allowing x to range over C. Thus

$$f(C) = \{ f(x) : x \in C \},$$

and $y \in f(C)$ means that $y = f(x)$ for some x belonging to C. The *image of f* is defined as $f(A)$.

Unfortunately, we cannot prove a theorem analogous to (1.5.4) for direct images, but there are a few reasonably nice properties.

1.5.5 Theorem.   *Let $f : A \to B$; then*
(a) *If $C \subseteq A$ then $C \subseteq f^{-1}\big[f(C)\big]$;*
(b) *If $D \subseteq B$ then $f\big[f^{-1}(D)\big] \subseteq D$.*
If we have an arbitrary collection of subsets A_i of A, then

(c) $f\left(\bigcup_i A_i \right) = \bigcup_i f(A_i)$ *and*

(d) $f\left(\bigcap_i A_i \right) \subseteq \bigcap_i f(A_i).$

Proof.
(a) If $x \in C$, then $f(x) \in f(C)$, in other words, $x \in f^{-1}\big[f(C)\big]$.
(b) If $y \in f\big[f^{-1}(D)\big]$, then $y = f(x)$ for some $x \in f^{-1}(D)$. But then $f(x) \in D$, and since $y = f(x)$ we have $y \in D$.
(c) We have

$$\begin{aligned}
y \in f\left(\bigcup_i A_i \right) \quad &\text{iff} \quad y = f(x) \text{ for some } x \in \bigcup_i A_i \\
&\text{iff} \quad \text{for at least one } i, \ y = f(x) \text{ for some } x \in A_i \\
&\text{iff} \quad \text{for at least one } i, \ y \in f(A_i) \\
&\text{iff} \quad y \in \bigcup_i f(A_i).
\end{aligned}$$

(d) If $y \in f(\bigcap_i A_i)$ then $y = f(x)$ for some $x \in \bigcap_i A_i$, and therefore y belongs to $f(A_i)$ for every i, that is, $y \in \bigcap_i f(A_i)$. $\blacksquare$

The inclusions in parts (a), (b), and (d) may be proper. In Figure 1.5.2 we have

$$f^{-1}\big[f(\{a_2\})\big] = f^{-1}(\{b_1\}) = \{a_1, a_2\},$$
$$f\big[f^{-1}\{b_1, b_2\}\big] = f(\{a_1, a_2\}) = \{b_1\},$$

and

$$f(\{a_1, a_3\} \cap \{a_2, a_3\}) = f(\{a_3\}) = \{b_3\},$$

but

$$f(\{a_1, a_3\}) \cap f(\{a_2, a_3\}) = \{b_1, b_3\} \cap \{b_1, b_3\} = \{b_1, b_3\}.$$

Problems For Section 1.5

1. Let $f(x) = x^3$ where we take the domain and codomain of f to be the set $\mathbb{R}$ of real numbers. Is f injective? surjective?

2. Same question for $f(x) = x^4$.

3. If $h(x) = (x^2 + 1)^{10}$, express h as the composition of two functions f and g.

4. Let $f : A \to B$ and suppose that $f(x)$ takes the same value c for every x in the domain of f. Can f ever be injective? surjective?

5. Let $f : A \to B$ where A has m elements and B has n elements (m and n are positive integers). If f is injective, show that $m \leq n$.

6. In Problem 5, if f is surjective, show that $m \geq n$.

7. . Let $f : A \to B$, and let C be a subset of A. If f is injective, show that $C = f^{-1}\big[f(C)\big]$.

8. Let $f : A \to B$, and let D be a subset of B. If f is surjective, show that $f\big[f^{-1}(D)\big] = D$.

9. Show that $f(\bigcap_i A_i) = \bigcap_i f(A_i)$ if f is injective.

1.6 Relations

If f is a function from A to B, then given any x in the domain A, the rule defining f produces an element $f(x)$ in the codomain B. There is no ambiguity; once x is specified, $f(x)$ is determined. In a mapping diagram such as Figure 1.5.2, each point of A has exactly one arrow leaving from it. This situation stimulates the mathematician's instinct to generalize. What would happen if we were to allow more than one arrow from some of the points of A? We would not have a function any more, but we would arrive at a new mathematical object called a *relation*. To define this formally, we need some terminology.

1.6.1 Definitions. An *ordered pair* (a, b) is (roughly) a set of two elements in which order counts, so that for example, $(3, 5)$ is not the same as $(5, 3)$. The key idea is that two ordered pairs (a, b) and (c, d) are regarded as identical if and only if $a = c$ and $b = d$. One way to capture this notion is to define an ordered pair as a function $f : \{1, 2\} \rightarrow \{a, b\}$ with $f(1) = a$ and $f(2) = b$. Similarly, an *ordered n-tuple* $(a_1, \ldots, a_n)$ may be defined as a function $f : \{1, 2, \ldots, n\} \rightarrow \{a_1, \ldots, a_n\}$ with $f(i) = a_i$, $1 \leq i \leq n$. But it is probably best not to worry about formal definitions and instead concentrate on the key idea: two ordered n-tuples $(a_1, \ldots, a_n)$ and $(b_1, \ldots, b_n)$ are the same if and only if $a_i = b_i$ for all i.

If A and B are sets, the *Cartesian product* of A and B, written $A \times B$, is the set of all ordered pairs (a, b), $a \in A$, $b \in B$. Similarly, the cartesian product of n sets $A_1, \ldots, A_n$, written $A_1 \times \ldots \times A_n$, is the set of all ordered n-tuples $(a_1, \ldots, a_n)$, where $a_i \in A_i$, $1 \leq i \leq n$. If $A_i = A$ for all i, the cartesian product is often written as A^n.

A *relation* between two sets A and B is a subset of $A \times B$; a *relation on A* is a subset of $A \times A$. Similarly, an *n-ary relation* among sets $A_1, \ldots, A_n$ is a subset of $A_1 \times \ldots \times A_n$; an *n-ary relation on A* is a subset of A^n.

One sign of mathematical maturity is to be able to absorb formal definitions without much pain, and to generate concrete examples illustrating the definitions. But at this early stage, a detailed example will be very useful.

1.6.2 Example.   We define a relation R on the set $A = \{1, 2, 3, 4, 5, 6, 7, 8\}$ as follows: the ordered pair (a, b) will belong to R (sometimes denoted aRb) if and only if $a < b$ and a divides b. Here is a list of all the ordered pairs in R.

$$(1, 2), \ (1, 3), \ (1, 4), \ (1, 5), \ (1, 6), \ (1, 7), \ (1, 8), \ (2, 4), \ (2, 6), \ (2, 8), \ (3, 6), \ (4, 8)$$

Note that $(3, 3) \notin R$; although 3 divides 3, 3 is not less than 3.

You can see the reason for the term "relation". For (a, b) to belong to R, a must be related to b in a certain way, namely a must be less than b and must be a divisor of b as well.

A relation between A and B can be represented by a mapping diagram as in Figure 1.5.2, but in this case there can be more than one arrow leaving a point of A. For the relation R just given, there will be 7 arrows leaving the point 1 and 3 arrows leaving 2, but only a single arrow leaving 3 and 4.

We are going to concentrate on the two types of relations that are most important in all areas of mathematics: equivalence relations and partial orderings. Let's start with equivalence relations; as we develop the theory, we will carry along an explicit example.

1.6.3 Equivalence Relations.   Let $A = \mathbb{Z}$, the set of all integers, and define a relation R on A as follows:

$$aRb \text{ if and only if } a - b \text{ is divisible by } 4.$$

In this case we say that a is *congruent* to b *modulo* 4, written as $a \equiv b$ mod 4. In a similar fashion we may define the relation of congruence modulo m for any positive integer $m \geq 2$. (Negative integers can be used, but do not contribute anything significant since $a - b$ if divisible by $-m$ if and only if it is divisible by m. The case $m = 1$ is legal but uninteresting because any two integers are congruent modulo 1.)

Now $a - b$ is divisible by 4 if and only if a and b leave the same remainder when divided by 4. For example, if $a = -5$ and $b = 7$, then aRb since $a - b = -12 = 4(-3)$.

Also, $a = 4(-2) + 3$ and $b = 4(1) + 3$, so both a and b leave remainder 3. In general, if $a = 4s + i$ and $b = 4t + i$, then $a - b = 4(s - t)$, a multiple of 4. Conversely, if $a - b = 4s$ and $b = 4t + i$, then $a = 4(s + t) + i$, so both a and b leave remainder i. The relation R has the following properties.

1. R is *reflexive*: aRa for all $a \in A$ (since $a - a = 0$, which is divisible by 4);
2. R is *symmetric*: if aRb then bRa (if $a - b = 4s$ then $b - a = 4(-s)$);
3. R is *transitive*:   if aRb and bRc then aRc (if $a - b = 4s$ and $b - c = 4t$ then

$$a - c = (a - b) + (b - c) = 4(s + t)).$$

A relation that is reflexive, symmetric and transitive is called an *equivalence relation*. If $a \in A$, the set $S(a)$ of elements equivalent to a (that is, the set of all $b \in A$ such that bRa) is called the *equivalence class* of a.

Let's find the equivalence classes for our explicit example, congruence modulo 4. $S(0)$ is the set of integers that leave remainder 0 when divided by 4, so that $S(0)$ consists of all multiples of 4. Thus

$$S(0) = \{\ldots - 8, -4, 0, 4, 8, \ldots\}.$$

Similarly,

$$S(1) = \{\ldots - 7, -3, 1, 5, 9, \ldots\}$$
$$S(2) = \{\ldots - 6, -2, 2, 6, 10, \ldots\}$$
$$S(3) = \{\ldots - 5, -1, 3, 7, 11, \ldots\}.$$

Notice that the sets $S(i)$, $i = 0, 1, 2, 3$, form a *partition* of A, that is, they are disjoint and their union is A. But what about the sets $S(i)$ when i is not 0, 1, 2 or 3? We don't get anything new. For example, $S(5)$ is the set of all elements equivalent to 5, i.e., the set all elements that leave remainder 1 when divided by 4. Thus $S(5)$ coincides with $S(1)$. In general, if $b \in S(a)$, then $S(b) = S(a)$, as we will verify in a moment.

Let's now prove that if R is any equivalence relation on A, then the equivalence classes form a partition of A. For any $a \in A$ we have $a \in S(a)$ by reflexivity, so that the union of the equivalence classes is A. Now consider two equivalence classes $S(a)$ and $S(b)$. There are two cases:

Case 1. b is equivalent to a, so that $b \in S(a)$. Then $S(b) = S(a)$. For if c belongs to $S(b)$, then cRb; but bRa, so that cRa by transitivity, and $c \in S(a)$. Conversely, if $c \in S(a)$ then cRa; but bRa, so that aRb by symmetry, and therefore cRb by transitivity, proving that $c \in S(b)$.

Case 2. b is not equivalent to a, so that $b \notin S(a)$. Then $S(b) \cap S(a) = \varnothing$. For if c belongs to both $S(b)$ and $S(a)$, then cRb and cRa; but then bRc by symmetry, hence bRa by transitivity. Thus $b \in S(a)$, which is a contradiction.

We have shown that the equivalence classes are disjoint sets whose union is A, as required.

You may not have seen congruence modulo m previously, but there is one equivalence relation that is extremely familiar. Let R be the *equality relation* on A, that is, aRb if and only if $a = b$. Equality is reflexive ($a = a$), symmetric (if $a = b$ then $b = a$), and transitive (if $a = b$ and $b = c$ then $a = c$). The equivalence class of the element $a \in A$ is the set $\{a\}$ consisting of a alone.

We now turn to partial orderings, and again we will carry along a concrete example.

1.6.4 Partial Orderings. Let $A = \{1, 2, \ldots, 12\}$, the set consisting of the first 12 positive integers. We define a relation R on A by aRb if and only if a divides b (often written as $a|b$). The relation R has the following properties:

R is *reflexive*: aRa for all $a \in A$ (since a divides a);

R is *antisymmetric*: if aRb and bRa then $a = b$ (if a divides b and b divides a, then in particular, $a \leq b$ and $b \leq a$, so that $a = b$);

R is *transitive*:   if aRb and bRc then aRc (if b is a multiple of a and c is a multiple of b, then c is a multiple of a).

A relation that is reflexive, antisymmetric and transitive is called a *partial ordering*. As with equivalence relations, there is a particular partial ordering that is very familiar. If A is any set of integers (or equally well, any set of real numbers), let aRb if and only if $a \leq b$. Then R is reflexive ($a \leq a$), antisymmetric (if $a \leq b$ and $b \leq a$ then $a = b$), and transitive (if $a \leq b$ and $b \leq c$ then $a \leq c$). But there is an important difference between the divisibility relation and the relation $\leq$. The latter relation is *total* in the sense that any two elements a and b can be compared: either $a \leq b$ or $b \leq a$. But it is not true that either a divides b or b divides a. For example, take $a = 4$ and $b = 7$.

Although the divisibility relation is not total on $A = \{1, 2, \ldots, 12\}$, it is total on various subsets of A. For example, let $B = \{1, 3, 6, 12\}$. Then any two elements of B can be compared; B is called a *totally ordered subset* or a *chain* of A.

If R is a partial ordering on a set A, we say that A is a *partially ordered set* under the relation R or that A is *partially ordered* by R. If in addition R is total, we say that R is a *total ordering* and that A is a *totally ordered set* under R (or that A is *totally ordered* by R).

Later we will be searching for *maximal elements* in a partially ordered set. If R is a partial ordering on A, the element $a \in A$ is said to be *maximal* if there is nothing bigger (with respect to R); in other words, there is no element b in A such that aRb and $a \neq b$. Equivalently, if $b \in A$ and aRb, then $a = b$. If R is the divisibility relation on $A = \{1, 2, \ldots, 12\}$, then 12 is definitely a maximal element, but there are several others, namely 7, 8, 9, 10 and 11. For example, 7 is maximal because there is no other element *in* A which is a multiple of 7.

The following theorem illustrates a basic technique involving finite sets.

1.6.5 Theorem. *If A is a nonempty finite set that is partially ordered by the relation R, then A has at least one maximal element.*

Proof. Pick any $a \in A$. If a is maximal, we are finished. If not, there is an element $b \in A$ with aRb and $a \neq b$. If b is maximal, we are finished. If not, there is an element $c \in A$ with bRc and $b \neq c$. In fact c cannot equal a, for if so, then we have aRb and bRa

(because bRc and $c = a$), hence $a = b$, a contradiction. Thus a, b, and c are distinct. If c is maximal, we are finished. If not, there is an element $d \in A$ with cRd and $c \neq d$, and in fact a, b, c, and d are distinct. If, for example, $d = a$, then aRc (by transitivity) and cRa (because cRd and $d = a$). Thus $a = c$, a contradiction. If we continue this procedure, then since A is finite, we will eventually run out of elements; in other words, the process will terminate at a maximal element. ∎

The above proof is an example of an *inductive procedure*. At step n of the procedure, we generate $a_n \in A$, where the element a_{n+1} depends on the results obtained at previous stages. If A has no maximal elements, we can construct an infinite chain of distinct elements a_n such that $a_n R a_{n+1}$ for every n. This is certainly impossible if A is finite.

Problems For Section 1.6

1. Let A be the set of all 3-letter "words" xyz, where x, y and z are English letters $a, b, \ldots, z$. Define a relation R on A by xRy if and only if x and y begin with the same letter. Show that R is an equivalence relation. How many equivalence classes are there?

2. Let R be an equivalence relation, so that in particular, R is symmetric. If in addition, R is antisymmetric, show that R is the equality relation.

3. When analyzing a theorem, mathematicians like to check if hypotheses can be weakened or dropped. In Problem 2, we assumed that R is an equivalence relation that is also antisymmetric. Suppose we make the weaker assumption that R is any relation that is symmetric and antisymmetric, and try to prove that R is the equality relation. Show that the argument given in the solution to Problem 2 fails.

4. (Continuing Problem 3) If a particular argument fails, this does not mean that a result is invalid. But in this case, an *explicit example* can be given of a relation R that is symmetric and antisymmetric, but not the equality relation. Supply such an example.

5. (Continuing Problem 4) Replace the hypothesis that R is an antisymmetric equivalence relation by an appropriately weaker hypothesis.

Attention. It is important to understand what "weaker" means in the above discussion. The hypothesis H_2 is said to be *weaker* than the hypothesis H_1 if H_1 implies H_2. In this case we also say that H_1 is *stronger* than H_2, and that H_2 is *more general* than H_1. Thus in Problem 3 we have H_1: R is an antisymmetric equivalence relation; and H_2: R is a relation that is symmetric and antisymmetric.

6. Let R be a partial ordering with more than one maximal element. Can R be total?

7. The *inclusion relation* on the collection of subsets of a set W is defined by ARB if and only if $A \subseteq B$. Show that the inclusion relation is a partial ordering. Is it total?

8. Suppose that we have an arbitrary collection of subsets A_i of a set W, and that B is the union of the A_i.

(a) Show that B is an *upper bound* of the A_i, that is, $A_i \subseteq B$ for every i.

(b) Show that if C is any upper bound of the A_i, then $B \subseteq C$; we say that B is the *least upper bound* of the A_i.

9. Suppose that we have an arbitrary collection of subsets A_i of a set W, and that B is the intersection of the A_i.

(a) Show that B is a *lower bound* of the A_i, that is, $B \subseteq A_i$ for every i.

(b) Show that if C is any lower bound of the A_i, then $C \subseteq B$; we say that B is the *greatest lower bound* of the A_i.

2

Counting

2.1 Fundamentals

The area of mathematics that deals with finding or estimating the size of sets is called combinatorics. If a set is small enough, its size can be found by simply counting the elements. But when a set is too large for this, techniques of counting must be developed. At an advanced level, combinatorial problems tend to be solved by diabolically clever tricks rather than the application of broad general principles, but we will only need some very basic results, which can be analyzed quite systematically. We start with the most fundamental principle of all.

2.1.1 Multiplication Rule. If we select an ordered n-tuple $(a_1, \ldots, a_n)$, where each a_i belongs to a set with k_i elements, then the total number of selections is $k_1 k_2 \cdots k_n$.

The case $n = 2$ is most familiar, and in this case the rule says that if a choice can be made in k_1 ways, and for each of these a second choice can be made in k_2 ways, then there are $k_1 k_2$ ways of making the pair of selections.

For example, if a computer password consists of an English letter (26 choices: A, B, ..., Z) followed by a single digit (10 choices: 0, 1, ..., 9), the number of possible passwords is $(26)(10) = 260$. To verify this, note that there are 10 passwords beginning with A ($A0, A1, \ldots, A9$), 10 passwords beginning with B, etc. Since there are 26 letters, we get $(26)(10)$ passwords altogether. In general, there are k_2 ordered pairs (a_1, a_2) beginning with a_1, and there are k_1 choices for a_1, so there are $k_1 k_2$ ordered pairs (a_1, a_2). For each such (a_1, a_2), there are k_3 ordered triples (a_1, a_2, a_3), which gives $k_1 k_2 k_3$ possible ordered triples (a_1, a_2, a_3). ("Ordered triple" is usually preferred to "ordered 3-tuple".) Continuing in this fashion (formally, we would use mathematical induction), we get $k_1 k_2 \cdots k_n$ ordered n-tuples altogether.

The Multiplication Rule covers selections in which order counts, but there are many situation where order does not matter. For example, suppose that we select two distinct integers from the the set $\{1, 2, 3, 4, 5\}$, without regard to order, so that 23 is the same as 32. (23 and 32 are abbreviations for the two-element subset $\{2, 3\}$.) The Multiplication Rule gives 25 ordered pairs (a, b) with a and b belonging to $\{1, 2, 3, 4, 5\}$, but there are only 10 two-element subsets, namely 12, 13, 14, 15, 23, 24, 25, 34, 35, 45. Although the

Multiplication Rule does not directly address this problem, it will still be helpful, as we will see shortly.

There are many problems in which repetition is not allowed, in other words, if an element is selected, it cannot appear again. For example, suppose we pick an ordered triple (a, b, c), with $a, b, c \in \{1, 2, 3, 4, 5\}$, without repetition. We can choose a in five ways, but having picked a, there are only four choices for b, and then only three choices for c. The total number of outcomes is, by the Multiplication Rule, $(5)(4)(3) = 60$.

Here is a very general combinatorial problem. Start with a set with n elements; for convenience let's use $\{1, 2, \ldots, n\}$. We are going to select k elements under various constraints; borrowing some terminology from statistics, we will call the result a *sample of size k* (*out of n*). Either order will count or it won't, and repetition (also referred to as *replacement*) will be allowed or it won't. This gives us four possible problems, which we analyze in turn.

2.1.2 Ordered Samples With Replacement.

Here we are selecting an ordered k-tuple $(a_1, \ldots, a_k)$, with repetition allowed and each a_i belonging to $\{1, 2, \ldots, n\}$. The number of possible selections is, by the Multiplication Rule,

$$(n)(n) \cdots (n) \ (k \text{ factors}) = n^k.$$

It is often convenient to visualize the selection process as the filling of k consecutive slots, with n possible choices for the contents of each slot; we get $(n)(n) \cdots (n) = n^k$ as above.

2.1.3 Ordered Samples Without Replacement.

Here we select an ordered k-tuple $(a_1, \ldots, a_k)$ as above, but once a number appears in the sequence, it can't be used again. Thus there are n choices at step 1, $n - 1$ choices at step 2, $n - 2$ at step 3, $\ldots, n - k + 1$ at step k. The number of possible selections is $(n)(n - 1)(n - 2) \cdots (n - k + 1)$, which can be written as

$$\frac{n!}{(n - k)!}$$

if $k < n$. If $k = n$, we are simply rearranging the integers in the sequence $12 \ldots n$; such a rearrangement is called a *permutation* of $\{1, 2, \ldots, n\}$. The number of permutations is $(n)(n - 1)(n - 2) \ldots (1) = n!$. Thus the above formula remains valid if we define 0! to be 1. For example, there are $3! = 6$ permutations of $\{1, 2, 3\}$, namely,

$$123, \ 132, \ 213, \ 231, \ 312, \ 321.$$

A permutation may be regarded as a bijective mapping from a set to itself. For example, the permutation 213 of $\{1, 2, 3\}$ corresponds to the function given by $f(1) = 2$, $f(2) = 1$, $f(3) = 3$.

2.1.4 Unordered Samples Without Replacement.

In this case we are selecting k distinct elements from $\{1, 2, \ldots, n\}$, without regard to order; in other words, we are forming a k-element subset of a set with n elements. The number of possible outcomes is denoted by $\binom{n}{k}$, and called the number of *combinations* of k objects out of n. We considered the case $n = 5$, $k = 2$ above, and we found that $\binom{5}{2} = 10$. To find $\binom{n}{k}$, we construct a two-stage process:

1. Select k distinct numbers from $\{1, \ldots, n\}$, without regard to order.
2. Order the selection made in step 1.

For example, if $n = 5$, $k = 3$, a possible selection at step 1 is 124 $(= \{1, 2, 4\})$, and this gives rise to $3! = 6$ possible outcomes at step 2, namely,

$$124, \ 142, \ 214, \ 241, \ 412, \ 421.$$

In general, there are $\binom{n}{k}$ possible choices at step 1, and for each of these results, there are $k!$ orderings at step 2. But after we have completed step 2, we have an ordered sample of size k out of n, without replacement, and we know that there are $\frac{n!}{(n-k)!}$ of these. By the Multiplication Rule,

$$\binom{n}{k} k! = \frac{n!}{(n-k)!},$$

that is,

$$\binom{n}{k} = \frac{n!}{k!(n-k)!}.$$

The application of the Multiplication Rule requires some elaboration. The product $\binom{n}{k} k!$ counts ordered pairs such as $(\{1, 2, 4\}, 124)$; $(\{1, 2, 4\}, 142)$; $(\{1, 2, 4\}, 214)$; $(\{1, 2, 4\}, 241)$; $(\{1, 2, 4\}, 412)$; and $(\{1, 2, 4\}, 421)$. On the other hand, $\frac{n!}{(n-k)!}$ counts ordered samples such as 124, 142, 214, 241, 412 and 421. But the two lists are in one-to-one correspondence because a sequence such as 412 immediately determines the underlying set $\{1, 2, 4\}$.

It is convenient to define $\binom{n}{0} = 1$; then $\binom{n}{k} = \binom{n}{n-k}$ for $k = 0, 1, \ldots, n$. This can also be seen by combinatorial reasoning. For example, selecting four distinct objects from a set of seven is equivalent to selecting three objects to be omitted.

2.1.5 Unordered Samples With Replacement ("Stars and Bars"). Now we are writing down k numbers from $\{1, \ldots, n\}$ with repetition allowed, but not paying attention to order. For example, suppose $n = 3$ and $k = 4$. The possibilities are 1111, 1112, 1113, 1122, 1123, 1133, 1222, 1223, 1233, 1333, 2222, 2223, 2233, 2333, 3333. Notice that 1213 gives nothing new because order does not count; 1213 is identified with 1123. Combinatorial problems are often solved by looking at them in just the right way. In this case, a sample is determined by the number of occurrences of 1, 2 and 3. If x_i is the number of times the integer i occurs in the sample, we have $x_1 + x_2 + x_3 = 4$. For example, 1133 corresponds to $x_1 = 2$, $x_2 = 0$, $x_3 = 2$. In general, we are looking for the number of solutions of the equation $x_1 + \cdots + x_n = k$, where the x_i are nonnegative integers. Each solution can be represented by a sequence of stars and bars, as follows:

$$|\ |\ x_1 \text{ stars } |\ x_2 \text{ stars } | \cdots | \ x_n \text{ stars } |\ |.$$

For the above example with $n = 3$, $k = 4$, $x_1 = 2$, $x_2 = 0$, $x_3 = 2$, we have

$$|\ |\ * \ * \ |\ \ |\ * \ * \ |\ |.$$

Similarly, $| \, | \, * \, | \, * \, * \, | \, * \, | \, |$ corresponds to $x_1 = 1$, $x_2 = 2$, $x_3 = 1$. With the double bars on the left and right always present, we are arranging k stars and $n - 1$ bars (forming n boxes for the stars) in a row. An arrangement is determined by selecting k positions out of $k + n - 1$ for the stars. The total number of arrangements, which is also the number of nonnegative integer solutions of $x_1 + \cdots + x_n = k$, is

$$\binom{k + n - 1}{k}.$$

When $n = 3$ and $k = 4$, we have $\binom{k+n-1}{k} = \binom{6}{4} = 15$, as found above by direct counting.

A permutation of a set $\{a_1, \ldots, a_n\}$ is an ordered n-tuple (also called an *ordering*) in which each a_i occurs exactly once. Thus if $n = 3$, then $a_3 a_1 a_2$ is a permutation, but $a_3 a_2 a_2$ is not. It is natural to generalize by allowing symbols to appear more than once. For example, AAABBCDD is an ordering in which A occurs 3 times, B and D twice, and C once. Other such orderings are ABCDADBA and DCAABABD. How many orderings are there altogether? The problem is solved by *position selection,* a basic combinatorial technique which has already occurred in (2.1.5). We are writing down a sequence of length 8, and the sequence is determined by selecting three positions out of eight for the A's, then two positions from the remaining five for the B's, then one position from the remaining three for the C, and finally two positions from the remaining two (only one choice!) for the D's. The Multiplication Rule gives

$$\binom{8}{3}\binom{5}{2}\binom{3}{1}\binom{2}{2}.$$

Using the formula

$$\binom{n}{k} = \frac{n!}{k!(n-k)!}$$

(see (2.1.4)), we obtain

$$\frac{8!}{3!2!1!2!}.$$

An ordering in which symbols are allowed to occur more than once is called a *generalized permutation.* Here is the general statement, which is proved by exactly the same reasoning as in the concrete example above.

2.1.6 Generalized Permutations.   Let $a_1, \ldots, a_r$ be distinct symbols, and let $k_1, \ldots, k_r$ be nonnegative integers whose sum is n. The number of ordered n-tuples in which a_i occurs exactly k_i times for all $i = 1, \ldots, r$ is $\frac{n!}{k_1! k_2! \cdots k_r!}$.

When $r = n$ and $k_i = 1$ for all i, the generalized permutation becomes an ordinary permutation, and $\frac{n!}{k_1! \cdots k_r!} = n!$, in agreement with (2.1.3). Also, when $r = 2$, $k_1 = k$ and $k_2 = n - k$, we are simply selecting k distinct positions out of n for a_1, and the positions for a_2 are then determined. The number of selections is $\frac{n!}{k!(n-k)!} = \binom{n}{k}$, in agreement with (2.1.4).

2.1.7 Examples.   Here are some representative counting problems.

1. Consider strings of length 7 formed from the digits $0, 1, \ldots, 9$. How many strings have exactly one 3 and exactly two 4's?

Pick a position for the 3, then two positions from the remaining six for the 4's, and then fill the remaining four slots with digits from $\{0, 1, 2, 5, 6, 7, 8, 9\}$. By the Multiplication Rule, the total number of acceptable strings is

$$7\binom{6}{2}(8^4).$$

2. From a pool of 50 people, choose five for a basketball game, four for a hockey game, and nine for a baseball game. How many choices are there if

(a) the three games are played simultaneously?

(b) the games are played on different days?

In part (a), a person can't be on more than one team, so the result is

$$\binom{50}{5}\binom{45}{4}\binom{41}{9}.$$

In part (b), a person can be on more than one team, so the number of choices is

$$\binom{50}{5}\binom{50}{4}\binom{50}{9}.$$

3. Consider six-letter words using only the letters P, Q, R. How many words have three of a kind and a pair, for example, QRQRQP?

Select the letter to appear three times, and then the letter to appear twice, and finally apply (2.1.6). The result is

$$3(2)\frac{6!}{3!2!1!}.$$

4. There are 6! ways to arrange $A_1, \ldots, A_6$ on a line. In how many ways can they be arranged on a circle?

Put A_1 down anywhere on the circle, then fill the seat adjacent to A_1 in the clockwise direction in five ways, then the seat two steps from A_1 in the clockwise direction in four ways, and so on. The number of circular permutations is 5!

5. Four companies C_1, C_2, C_3, C_4 bid for eleven different government grants. In how many ways can the grants be awarded if C_2 must get between two and five grants?

There are four mutually exclusive possibilities corresponding to C_2's getting 2, 3, 4 or 5 grants, and we can add the resulting answers. For example, if C_2 gets three grants (this can happen in $\binom{11}{3}$ ways), then each of the remaining eight grants can be awarded to C_1, C_3 or C_4. The result is

$$\binom{11}{2}3^9 + \binom{11}{3}3^8 + \binom{11}{4}3^7 + \binom{11}{5}3^6.$$

6. How many permutations of the 26 letters of the alphabet have exactly 3 letters between the A and the B?

Either A comes before B or vice versa (two choices). We can select the three letters to appear between A and B in $24(23)(22)$ ways, producing a block of five letters. The overall permutation is then determined by permuting the 5-block and the 21 remaining letters. Thus the answer is

$$2(24)(23)(22)(22!).$$

7. You have 10 hours to study for History, Math and Spanish. For example, two possibilities are (i) nine hours on History and one hour on Spanish and (ii) three hours on History, three hours on Math and four hours on Spanish. How many possibilities are there?

If x_1, x_2 and x_3 are the number of hours spent on History, Math and Spanish respectively, we are looking for nonnegative integer solutions of $x_1 + x_2 + x_3 = 10$. By (2.1.5), the number of possibilities is

$$\binom{10 + 3 - 1}{10} = \binom{12}{10} = 66.$$

8. Five people P_1, P_2, P_3, P_4 and P_5 sit down in a row containing 12 chairs. For example, one possibility is (with X standing for an empty chair)

$$\text{X } P_5 \text{ X } P_1 P_2 \text{ X } P_4 \text{ X X X } P_3 \text{ X}$$

Find the number of arrangements in which no two people are next to one another, in other words, all five people have elbow room.

We have seven identical empty chairs X X X X X X X, which determine two end spaces (to the left of the first X and to the right of the last X) and six interior spaces. P_1 goes in one of these eight spaces, and then P_2 goes in one of the remaining seven spaces, and so on. The result is

$$8(7)(6)(5)(4).$$

9. A poker hand with two pairs has, for example, two jacks, two kings and a fifth card that is not a jack or a king. To count the number of poker hands containing two pairs, suppose we do the following:

Step 1: Pick a face value (ace, king, queen, ..., 4, 3, 2) and then pick two cards having that face value.

Step 2. Pick another face value and then pick two cards having that face value.

Step 3. Pick the fifth card from the 44 not of the two face values chosen (to avoid a full house, e.g. three jacks and two kings).

The result is $13\binom{4}{2}(12)\binom{4}{2}(44)$. Why is this incorrect?

We are counting the following as different outcomes when they are really the same hand:

Outcome 1: Pick queen, hearts and spades; then pick jack, hearts and clubs; then pick the ace of clubs.

Outcome 2: Pick jack, hearts and clubs; then pick queen, hearts and spades; then pick the ace of clubs.

10. Do Problem 9 correctly.

Step 1: Pick two distinct face values out of 13 for the pairs.

Step 2: Pick two cards from each face value.

Step 3: Pick a fifth card from the 44 not of either face value.

The result is $\binom{13}{2}\binom{4}{2}\binom{4}{2}(44)$

The incorrect approach given in Problem 9 involves a **multiple count,** i.e., a process that counts some outcomes more than once. In fact, in this case each outcome is counted exactly twice, a perfect double count. Multiple counting is one of the most common mistakes made in solving combinatorial problems.

Problems For Section 2.1

1. Let A be a set with n elements. How many bijective functions are there from A to A?

2. Let A be a set with k elements, and B a set with n elements. How many functions are there from A to B?

3. In Problem 2, how many injective functions are there from A to B?

4. If A is a set with n elements, how many subsets does A have? (Suggestion: A subset may be identified with a function from A to $\{0, 1\}$. For example, if $A = \{1, 2, 3, 4, 5\}$, the subset $\{1, 3, 4\}$ corresponds to $f(1) = f(3) = f(4) = 1$, $f(2) = f(5) = 0$.)

5. (Labeled groupings). Suppose we have five people, call them 1, 2, 3, 4, 5. We send two people to room R_1, one person to room R_2, and two people to room R_3. In how many ways can this be done? (Note that each room is labeled by a positive integer. If we drop the labels, we change the problem to the partitioning of $\{1, 2, 3, 4, 5\}$ into subsets of sizes 2, 1 and 2, in other words two subsets of size 2 and one subset of size 1. The partitioning problem is considered in Problems 7,8 and 9.)

6. Generalizing Problem 5, if we have n people $1, 2, \ldots, n$, and we send k_i to room i, $i = 1, 2, \ldots, r$, where $k_1 + \cdots + k_r = n$, how many assignments are possible?

7. (Unlabeled groupings). Suppose we partition the set $\{1, 2, 3, 4, 5\}$ into two subsets of size 2 and one subset of size 1. The number of possible partitions is not the same as the number of assignments in Problem 5. Explain why, and compute the correct number of partitions.

8. Find the number of ways of partitioning a set with 24 elements into two subsets of size 5, four subsets of size 3, and one subset of size 2. (If we send five people to room R_1, five to R_2, three to R_3, three to R_4, three to R_5, three to R_6, and two to R_7, the number of assignments is

$$\frac{(24)!}{5!5!3!3!3!3!2!}.$$

But as in Problem 7, this must be divided by a correction factor to get the number of partitions.)

9. Generalizing Problem 8, suppose a set with n elements is partitioned into subsets of sizes $k_1, \ldots, k_r$. The k_i are nonnegative integers whose sum is n, but they are not necessarily distinct. Assume that among the k_i there are exactly t_j occurrences of u_j, $1 \leq j \leq m$, where the sum of the $t_j u_j$ is n. For example, in Problem 8 we have $k_1 = k_2 = 5$, $k_3 = k_4 = k_5 = k_6 = 3$, $k_7 = 2$; $t_1 = 2$, $u_1 = 5$, $t_2 = 4$, $u_2 = 3$, $t_3 = 1$, $u_3 = 2$. Find the number of possible partitions.

10. Find the number of possible orders of 10 bagels if there are five available types: plain, rye, poppy, garlic and cinnamon raisin.

2.2 The Binomial And Multinomial Theorems

The Binomial Theorem expresses $(a+b)^n$, the nth power of the binomial $a+b$, as a sum of terms of the form $c_k a^k b^{n-k}$, $k = 0, 1, \ldots, n$. For small values of n, we can do a direct computation:

$$(a+b)^0 = 1$$
$$(a+b)^1 = a+b$$
$$(a+b)^2 = a^2 + 2ab + b^2$$
$$(a+b)^3 = a^3 + 3a^2 b + 3ab^2 + b^3$$
$$(a+b)^4 = a^4 + 4a^3 b + 6a^2 b^2 + 4ab^3 + b^4.$$

The general formula may be derived by combinatorial reasoning. Let's illustrate the idea with $n = 5$; we have

$$(a+b)^5 = (a+b)(a+b)(a+b)(a+b)(a+b).$$

A typical term in the expansion of the right-hand side is generated by picking a letter (a or b) from each of the five factors $(a+b)$. For example, if we choose a, b, b, a, b in turn, we get $a^2 b^3$. There are other ways to generate $a^2 b^3$, e.g., a, a, b, b, b; b, b, a, b, a, etc. To count the number of selections that yield $a^2 b^3$, notice that we are selecting two positions out of five for the a's, and this will determine the positions of the b's. Thus (as in (2.1.6)) there are $\binom{5}{2} = 10$ occurrences of $a^2 b^3$. Similar reasoning gives the formula for $n = 5$:

$$(a+b)^5 = a^5 + 5a^4 b + 10a^3 b^2 + 10a^2 b^3 + 5ab^4 + b^5.$$

In general, $(a+b)^n = (a+b) \cdots (a+b)$, and there are $\binom{n}{k}$ selections that yield $a^k b^{n-k}$ in the expansion of the right-hand side. We have, therefore,

2.2.1 The Binomial Theorem.

$$(a+b)^n = \sum_{k=0}^{n} \binom{n}{k} a^k b^{n-k}.$$

In the formula, we haven't specified a and b. They can be arbitrary numbers (integers, rationals, real or complex numbers). They can even be more general mathematical objects such as polynomials; for example, $a = X^2 + 2X - 5$, $b = X + \pi$.

The Binomial Theorem can be generalized by replacing the binomial $a+b$ by a multinomial $a_1 + a_2 + \cdots + a_r$. Let's look at a particular case:

$$(a_1 + a_2 + a_3)^4 = (a_1 + a_2 + a_3)(a_1 + a_2 + a_3)(a_1 + a_2 + a_3)(a_1 + a_2 + a_3).$$

A typical term in the expansion of the right-hand side is $a_1 a_2 a_3^2$, and it can occur if a_1 is chosen from the first factor, a_2 from the second factor, and a_3 from the third and fourth factors. We abbreviate this sequence of choices by $a_1 a_2 a_3 a_3$. Using this notation, we can make a list of all possible ways of generating the term $a_1 a_2 a_3^2$:

$$a_1 a_2 a_3 a_3, \quad a_1 a_3 a_2 a_3, \quad a_1 a_3 a_3 a_2, \quad a_2 a_1 a_3 a_3, \quad a_2 a_3 a_1 a_3, \quad a_2 a_3 a_3 a_1,$$

$$a_3 a_1 a_2 a_3, \quad a_3 a_1 a_3 a_2, \quad a_3 a_2 a_1 a_3, \quad a_3 a_2 a_3 a_1, \quad a_3 a_3 a_1 a_2, \quad a_3 a_3 a_2 a_1.$$

In fact we can verify that there are 12 items on the list without writing out all possibilities. Notice that each sequence on the list is a generalized permutation of $a_1 a_2 a_3 a_3$, and by (2.1.6), the number of such permutations is $\frac{4!}{1!1!2!} = 12$. Similarly, the number of occurrences of $a_1^2 a_3^2$ is $\frac{4!}{2!2!} = 6$. In general,

$$(a_1 + \cdots + a_r)^n = (a_1 + \cdots + a_r) \cdots (a_1 + \cdots + a_r),$$

and the number of occurrences of $a_1^{k_1} \cdots a_r^{k_r}$, where the k_i are nonnegative integers summing to n, is $\frac{n!}{k_1! \cdots k_r!}$ by (2.1.6). Thus we have

2.2.2 The Multinomial Theorem.

$$(a_1 + \cdots + a_r)^n = \sum \left\{ \frac{n!}{k_1! \cdots k_r!} a_1^{k_1} \cdots a_r^{k_r} : k_1, \ldots, k_r \geq 0, \ k_1 + \cdots + k_r = n \right\}$$

The number of terms in the multinomial expansion is the number of nonnegative integer solutions of $k_1 + \cdots + k_r = n$, which is $\binom{n+r-1}{n}$ by (2.1.5). For the above example with $n = 4$ and $r = 3$, we have $\binom{4+3-1}{4} = 15$ terms, which we may list as follows.

k_1	k_2	k_3	$\dfrac{4!}{k_1! k_2! k_3!}$
4	0	0	1
0	4	0	1
0	0	4	1
3	1	0	4
3	0	1	4
0	3	1	4
0	1	3	4
1	0	3	4
1	3	0	4
2	2	0	6
2	0	2	6
0	2	2	6
2	1	1	12
1	2	1	12
1	1	2	12

The complete expansion is

$$(a_1 + a_2 + a_3)^4 = a_1^4 + a_2^4 + a_3^4 + 4a_1^3 a_2 + 4a_1^3 a_3 + 4a_2^3 a_3$$
$$+ 4a_2 a_3^3 + 4a_1 a_3^3 + 4a_1 a_2^3$$
$$+ 6a_1^2 a_2^2 + 6a_1^2 a_3^2 + 6a_2^2 a_3^2$$
$$+ 12a_1^2 a_2 a_3 + 12a_1 a_2^2 a_3 + 12a_1 a_2 a_3^2.$$

Problems For Section 2.2

1. Use the Binomial Theorem to show that $\sum_{k=0}^{n} \binom{n}{k} = 2^n$.

2. Use the Binomial Theorem to show that $\sum_{k=0}^{n} (-1)^k \binom{n}{k} = 0$.

3. Give another proof of the identity of Problem 1, based on Problem 4 of Section 2.1.

4. The binomial coefficients $\binom{n}{k}$ can be arranged in a form known as *Pascal's Triangle,* which looks like this:

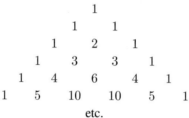

etc.

If we refer to the first row of the triangle as row 0, then row n consists of the coefficients $\binom{n}{k}$, $k = 0, 1, \ldots, n$. Notice that the element in row n, position k, is the sum of the two elements in row $n - 1$ that are diagonally above; in other words,

$$\binom{n}{k} = \binom{n-1}{k-1} + \binom{n-1}{k}.$$

Derive this identity by algebraic manipulation.

5. Derive the identity of Problem 4 by combinatorial reasoning.

6. In the expansion $(a_1 + a_2 + a_3)^4$ which we computed above, the sum of all the multinomial coefficients $\dfrac{4!}{k_1! k_2! k_3!}$ is $3 + 6(4) + 3(6) + 3(12) = 81$. State and prove a general result that would allow us to write down this number without computing all the terms of the expansion.

2.3 The Principle of Inclusion and Exclusion

If A and B are sets, it frequently happens that there is no direct way to compute $N(A \cup B)$, the number of elements in $A \cup B$. The Principle of Inclusion and Exclusion (abbreviated PIE) expresses $N(A \cup B)$ in terms of $N(A)$, $N(B)$ and $N(A \cap B)$, which are often easier to calculate. The formula states that

$$N(A \cup B) = N(A) + N(B) - N(A \cap B).$$

There is a similar formula for three sets A, B, C:

$$N(A \cup B \cup C) = N(A) + N(B) + N(C) - N(A \cap B) - N(A \cap C)$$
$$- N(B \cap C) + N(A \cap B \cap C).$$

To verify PIE for 2 and 3 sets, see Figure 2.3.1.

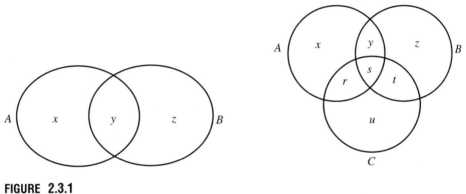

FIGURE 2.3.1
Illustrating PIE

The lower case letters stand for the number of elements in the indicated regions. For two sets A and B, we have

$$N(A \cup B) = x + y + z$$

and

$$N(A) + N(B) - N(A \cap B) = (x + y) + (y + z) - y = x + y + z$$

as asserted. For three sets A, B, C, we have

$$N(A \cup B \cup C) = x + y + z + r + s + t + u$$

and

$$N(A) + N(B) + N(C) - N(A \cap B) - N(A \cap C) - N(B \cap C) + N(A \cap B \cap C)$$
$$= (x + y + r + s) + (y + z + s + t) + (r + s + t + u)$$
$$- (y + s) - (r + s) - (s + t) + s$$
$$= x + y + z + r + s + t + u,$$

again as asserted. Notice that the term s is overcounted in $N(A) + N(B) + N(C)$. (It appears three times, each time with a $+$ sign.) But then it is subtracted three times in $-\big(N(A \cap B) + N(A \cap C) + N(B \cap C)\big)$, thus producing an undercount, and finally it is added back in $N(A \cap B \cap C)$, producing the correct count. The term "inclusion and exclusion" refers to this process of addition and subtraction.

There is a PIE formula for any finite number of sets. For four sets A, B, C, D we have

$$N(A \cup B \cup C \cup D) = N(A) + N(B) + N(C) + N(D) - N(A \cap B) - N(A \cap C)$$
$$- N(A \cap D) - N(B \cap C) - N(B \cap D) - N(C \cap D)$$
$$+ N(A \cap B \cap C) + N(A \cap B \cap D) + N(A \cap C \cap D)$$
$$+ N(B \cap C \cap D) - N(A \cap B \cap C \cap D).$$

Here is the general result:

2.3.1 Principle of Inclusion and Exclusion for n sets, abbreviated PIE$_n$.

$$N \left(\bigcup_{i=1}^{n} A_i \right) = \sum_{i=1}^{n} N(A_i) - \sum_{i<j} N(A_i \cap A_j) + \sum_{i<j<k} N(A_i \cap A_j \cap A_k)$$
$$- \cdots + (-1)^{n-1} N(A_1 \cap A_2 \cap \cdots \cap A_n).$$

Proof.   Venn diagrams will not help us for large n. A proof by mathematical induction is possible, but it is rather cumbersome. But there is a nice argument based on the overcounting and undercounting idea that we looked at in the $n = 3$ case. Any element in the union of the A_i will belong to some of the sets but not to others. Suppose that we consider an element x that belongs to exactly r of the sets, and for convenience in notation, let's assume that x belongs to $A_1, \ldots, A_r$ but not to $A_{r+1}, \ldots, A_n$. We must show that the PIE formula counts x exactly once.

Now x is counted r times in $\sum_{i=1}^{n} N(A_i)$; what about $\sum_{i<j} N(A_i \cap A_j)$? There are $\binom{n}{2}$ ways of selecting two distinct integers i and j (with $i < j$) out of n, but only $\binom{r}{2}$ of the terms $N(A_i \cap A_j)$ count x, since x belongs only to $A_1, \ldots, A_r$. Thus $\sum_{i<j} N(A_i \cap A_j)$ yields a count of $\binom{r}{2}$. By the same reasoning, $\sum_{i<j<k} N(A_i \cap A_j \cap A_k)$ yields a count of $\binom{r}{3}$, and so on. The overall count is

$$r - \binom{r}{2} + \binom{r}{3} - \cdots + (-1)^{r-1} \binom{r}{r}.$$

(We terminate at r, not n, because x belongs only to r of the sets, so cannot be counted by terms involving intersections of more than r sets.)

The proof of PIE has now been reduced to the verification of the identity

$$\sum_{k=1}^{r} (-1)^{k-1} \binom{r}{k} = 1.$$

But by the Binomial Theorem,

$$0 = (-1+1)^r = \sum_{k=0}^{r} \binom{r}{k} (-1)^k = 1 + \sum_{k=1}^{r} (-1)^k \binom{r}{k}.$$

Since $-(-1)^k = (-1)^{k-1}$, the result follows. ∎

2.3.2 Example.
There are 26^n potential English words of length n, that is, ordered n-tuples with components $A, B, \ldots, Z$. How many words contain at least one A?

We can use PIE, but it would be overkill. If A_i is the set of English words with an A in position i, we are looking for $N(A_1 \cup \cdots \cup A_n)$. Here is a much easier approach:

Number of words with at least one A

$$= \text{ total number of words } - \text{ number of words with no } A$$

$$= 26^n - 25^n$$

(since a word with no A is a word from a 25-letter alphabet).

If you have a problem involving "at least one", first try "total minus none" before bringing out the heavy PIE machinery. In the next example, we really do need PIE.

2.3.3 The Euler Phi Function. Two integers a and b are said to be *relatively prime* if they have no common factors except 1 and -1. We also say that a is *relatively prime to* b. Thus 8 and 15 are relatively prime, but 9 and 12 are not, since 3 is a common factor. The *Euler phi function* is defined for positive integers n by

$$\varphi(n) = \text{ the number of integers in } \{1, \ldots, n\} \text{ that are relatively prime to } n.$$

Since 1 is always relatively prime to n, we have $\varphi(n) \geq 1$ for all n. Since n cannot be relatively prime to n for $n > 1$, we can replace the set $\{1, \ldots, n\}$ by $\{1, \ldots, n-1\}$ if $n > 1$.

For small n, $\varphi(n)$ can be computed directly; here is a brief table.

n	1	2	3	4	5	6	7	8	9	10	11	12	13	14	15
$\varphi(n)$	1	1	2	2	4	2	6	4	6	4	10	4	12	6	8

We are going to use PIE to find an explicit formula for $\varphi(n)$. We must first factor n into primes (integers $p = 2, 3, 5, 7, 11, 13, 17, 19$, etc., whose only divisors are ± 1 and $\pm p$). For example, $n = 45000 = (2^3)(3^2)(5^4)$; in the next chapter we will show that the primes appearing in the factorization (2, 3 and 5 in this case) and the associated exponents (3, 2 and 4) are unique. Let's look at this numerical example in more detail. How can a positive integer m *not* be relatively prime to n? It must have a factor in common with n; in other words, m must be a multiple of 2, 3 or 5. This is the basis of our strategy.

Define A_1 to be the set of all integers in $\{1, \ldots, n\}$ that are multiples of 2, and similarly define A_2 to be the set of multiples of 3, and A_3 the set of multiples of 5. Then

$$\varphi(n) = n - N(A_1 \cup A_2 \cup A_3).$$

By the Principle of Inclusion and Exclusion,

$$\varphi(n) = n - N(A_1) - N(A_2) - N(A_3) + N(A_1 \cap A_2)$$
$$+ N(A_1 \cap A_3) + N(A_2 \cap A_3) - N(A_1 \cap A_2 \cap A_3).$$

If p divides n, then there are exactly n/p multiples of p in $\{1, \ldots, n\}$. Thus,

$$N(A_1) = \frac{n}{2}, \quad N(A_2) = \frac{n}{3}, \quad N(A_3) = \frac{n}{5}.$$

Since $A_1 \cap A_2$ consists of all multiples of $(2)(3) = 6$ between 1 and n,

$$N(A_1 \cap A_2) = \frac{n}{(2)(3)};$$

similarly,

$$N(A_1 \cap A_3) = \frac{n}{(2)(5)}, \quad N(A_2 \cap A_3) = \frac{n}{(3)(5)},$$

and

$$N(A_1 \cap A_2 \cap A_3) = \frac{n}{(2)(3)(5)}.$$

Therefore,

$$\varphi(n) = n\left(1 - \frac{1}{2} - \frac{1}{3} - \frac{1}{5} + \frac{1}{(2)(3)} + \frac{1}{(2)(5)} + \frac{1}{(3)(5)} - \frac{1}{(2)(3)(5)}\right)$$
$$= n\left(1 - \frac{1}{2}\right)\left(1 - \frac{1}{3}\right)\left(1 - \frac{1}{5}\right).$$

The general result can be predicted from this pattern.

2.3.4 Theorem. *If the prime factorization of n is $n = p_1^{e_1} p_2^{e_2} \cdots p_k^{e_k}$, then*

$$\varphi(n) = n\left(1 - \frac{1}{p_1}\right)\left(1 - \frac{1}{p_2}\right)\cdots\left(1 - \frac{1}{p_k}\right).$$

In particular, if p is prime, then $\varphi(p^e) = p^e\left(1 - \frac{1}{p}\right) = p^e - p^{e-1}$, so $\varphi(p) = p - 1$.

Proof. Let A_i be the set of integers in $\{1, \ldots, n\}$ that are multiples of p_i. Then

$$\varphi(n) = n - N(A_1 \cup A_2 \cup \cdots \cup A_k)$$
$$= n - N(A_1) - N(A_2) - \cdots - N(A_k) + N(A_1 \cap A_2) + N(A_1 \cap A_3)$$
$$+ \cdots + N(A_{k-1} \cap A_k) - \cdots + (-1)^k N(A_1 \cap A_2 \cap \cdots \cap A_k)$$
$$= n\left(1 - \frac{1}{p_1} - \frac{1}{p_2} - \cdots - \frac{1}{p_k} + \frac{1}{p_1 p_2} + \frac{1}{p_1 p_3}\right.$$
$$\left. + \cdots + \frac{1}{p_{k-1} p_k} - \cdots + (-1)^k \frac{1}{p_1 p_2 \cdots p_k}\right)$$
$$= n\left(1 - \frac{1}{p_1}\right)\left(1 - \frac{1}{p_2}\right)\cdots\left(1 - \frac{1}{p_k}\right). \blacksquare$$

Problems For Section 2.3

In this problem set, we will apply the Principle of Inclusion and Exclusion to solve two famous problems.

A *derangement* of $\{1, 2, \ldots, n\}$ is a permutation in which no integer stands in its natural position. Thus if $n = 4$, the derangements are 2143, 2341, 2413, 3142, 3412, 3421, 4123, 4312, and 4321; for example, 3214 is not a derangement because 2 is in position 2; 1432 is not a derangement because 1 and 3 are in positions 1 and 3, respectively.

1. Let A_i be the set of permutations of $\{1, \ldots, n\}$ in which the integer i stands in position i. Show that $N(A_i) = (n-1)!$

2. If $i_1, \ldots, i_k$ are distinct integers in $\{1, \ldots, n\}$, show that

$$N\left(A_{i_1} \cap \cdots \cap A_{i_k}\right) = (n - k)!$$

3. Show that the number $d(n)$ of derangements of $\{1, 2, \ldots, n\}$ is

$$\sum_{i=0}^{n} (-1)^i \binom{n}{i} (n - i)!,$$

which can also be expressed as

$$n! \sum_{i=0}^{n} \frac{(-1)^i}{i!}.$$

4. (Uses calculus). Recall the series expansion

$$e^{-1} = \sum_{i=0}^{\infty} \frac{(-1)^i}{i!},$$

and show that $d(n)$ is the nearest integer to $n!/e$.

In Section 2.1, Problem 3, we computed the number of injective functions from a set A with k elements to a set B with n elements. We now have enough machinery to compute the number of surjective functions.

5. Let A_i be the number of functions f from $\{1, \ldots, k\}$ to $\{1, \ldots, n\}$ such that i does not belong to the image of f; in other words, $f(x)$ is never equal to i. Show that $N(A_i) = (n - 1)^k$.

6. If $i_1, \ldots, i_r$ are distinct integers in $\{1, \ldots, n\}$, show that

$$N\left(A_{i_1} \cap \cdots \cap A_{i_r}\right) = (n - r)^k.$$

7. Show that the number of surjective function from $\{1, \ldots, k\}$ to $\{1, \ldots, n\}$ is

$$\sum_{i=0}^{n} (-1)^i \binom{n}{i} (n - i)^k.$$

8. Consider a partition of $\{1, 2, 3, 4, 5, 6, 7, 8\}$ into four disjoint nonempty subsets, for example, $\{1, 2\}$, $\{3, 4, 5\}$, $\{6\}$ and $\{7, 8\}$. Such a partition gives rise to a surjective function $f : \{1, 2, 3, 4, 5, 6, 7, 8\} \rightarrow \{1, 2, 3, 4\}$, namely,

$$f(1) = f(2) = 1, \qquad f(3) = f(4) = f(5) = 2,$$
$$f(6) = 3, \qquad f(7) = f(8) = 4.$$

Conversely, given a surjective function such as f, we can immediately recover the corresponding partition. It follows that the number of surjective functions from a set of size 8 to a set of size 4 is the same as the number of partitions of a set of size 8 into four disjoint nonempty subsets.

Find the flaw in this argument, and give the correct result.

9. Express the number $S(k, n)$ of surjective functions from $\{1, \ldots, k\}$ to $\{1, \ldots, n\}$ in terms of the number of partitions $P(k, n)$ of $\{1, \ldots, k\}$ into n disjoint nonempty subsets.

10. Compute $S(k, n)$ and $P(k, n)$ when $k = 4$ and $n = 3$. List all the partitions of $\{1, 2, 3, 4\}$ into three disjoint nonempty subsets.

2.4 Counting Infinite Sets

What do we mean by the size of a set? For finite sets, there is no problem. If A is empty, its size is 0, and if A is nonempty and finite, then A is in one-to-one correspondence with $\{1, \ldots, n\}$ for some positive integer n, and we can confidently say that its size is n. But for infinite sets, there are some difficulties. The positive integers form a proper subset of the set of all integers, but are there more integers than there are positive integers? It seems more reasonable to say that the two sets have the same size, because there is a one-to-one correspondence between them, namely,

$$1 \leftrightarrow 0, \ 2 \leftrightarrow 1, \ 3 \leftrightarrow -1, \ 4 \leftrightarrow 2, \ 5 \leftrightarrow -2, \ 6 \leftrightarrow 3, \ 7 \leftrightarrow -3, \ldots$$

The set of all integers can be counted, that is, listed in the form $a_1, a_2, \ldots$. The above list (there are many others) is $0, 1, -1, 2, -2, 3, -3, \ldots$. (Notice that the counting scheme $0, 1, 2, 3, 4, \ldots, -1, -2, -3, -4, \ldots$ doesn't work; a machine that is programmed to print out the list will never reach the negative integers.)

2.4.1 Definitions. A set is *countably infinite* if it can be placed in one-to-one correspondence with the positive integers. A set is *countable* if it is finite or countably infinite, *uncountable* if it is not countable. *Uncountably infinite* is synonymous with *uncountable*.

The set $\mathbb{Z}^+$ of positive integers is countable by definition, and the set $\mathbb{Z}$ of all integers is countable, as we have just seen. In fact, the set $\mathbb{Q}$ of all rational numbers is countable as well; here is Cantor's classic argument. We arrange the positive rationals as follows:

$$
\begin{array}{ccccc}
1/1 & 1/2 & 1/3 & 1/4 & \ldots \\
2/1 & 2/2 & 2/3 & 2/4 & \ldots \\
3/1 & 3/2 & 3/3 & 3/4 & \ldots \\
4/1 & 4/2 & 4/3 & 4/4 & \ldots \\
\vdots & & & &
\end{array}
$$

We count the positive rationals by traversing the diagonals of the above array (and thus the method is known as *Cantor's Diagonal Process*). Explicitly, the list is

$$1/1, \ 2/1, \ 1/2, \ 3/1, \ 2/2, \ 1/3, \ 4/1, \ 3/2, \ 2/3, \ 1/4, \ldots$$

In general, we move northeast from $m/1$ to $1/m$, for each $m = 1, 2, \ldots$. It is true that every positive rational appears infinitely often on the list (for example, $2/1 = 4/2 = 6/3 = \ldots$), but this is not a problem since duplicates can simply be discarded. This scheme counts the positive rationals, and the technique that allowed us to proceed from positive integers to all integers will work here as well. We conclude that the set of all rationals is countable.

Cantor also proved that the set $\mathbb{R}$ of real numbers is uncountable, so it is reasonable to say that $\mathbb{R}$ is strictly larger than $\mathbb{Q}$ (or $\mathbb{Z}$, or $\mathbb{Z}^+$). Suppose that the reals between 0 and 1 can be listed as $x_1, x_2, x_3, \ldots$. If we were to write out the decimal expansion of each x_j, the list would look like this:

$$x_1 = .r_{11}r_{12}r_{13} \cdots$$

$$x_2 = .r_{21}r_{22}r_{23} \cdots$$

$$x_3 = .r_{31}r_{32}r_{33} \cdots$$

$$\vdots$$

We now form a real number $x = .r_1 r_2 r_3 \ldots$ by specifying that $r_1 \neq r_{11}$, $r_2 \neq r_{22}$, $r_3 \neq r_{33}$, and so on. (This is a slightly different diagonal argument; here we go down the main diagonal, forcing a mismatch at each step.) The number x is between 0 and 1, but it cannot appear on the list, since for every n it differs from x_n in the nth digit. (An objection can be raised at this point; what if $x_1 = .3000\ldots$, and x turns out to be $.2999\ldots$, the same number? We can avoid this difficulty by requiring that $1 \leq r_j \leq 8$ for all j.)

We will see in Chapter 4 that for any set A, the set of all subsets of A is strictly larger than A. At this point, after a brief digression on binary representation, we can say something about the case $A = \mathbb{Z}^+$, the set of positive integers.

Just as the decimal 86.357 represents the number

$$8(10)^1 + 6(10)^0 + 3(1/10) + 5(1/10)^2 + 7(1/10)^3,$$

the binary expression 110.1101 represents the number

$$1(2^2) + 1(2^1) + 0(2^0) + 1(1/2)^1 + 1(1/2)^2 + 0(1/2)^3 + 1(1/2)^4 = 6 + (13/16).$$

Every real number x has a binary expansion. For example, if $x = 1/3$, then

$$
\begin{array}{lll}
0 \leq x < 1/2 & \text{so} & x = .0\ldots \\
1/4 \leq x < 2/4 & \text{so} & x = .01\ldots \\
2/8 \leq x < 3/8 & \text{so} & x = .010\ldots \\
5/16 \leq x < 6/16 & \text{so} & x = .0101\ldots
\end{array}
$$

and so on. In fact if you perform ordinary long division of 1 by 11 (in binary), you will find that the pattern 01 repeats, in other words,

$$x = .01010101\ldots$$

To check that this is correct, note that

$$
\begin{aligned}
.01010101\ldots &= .01 + .0001 + .000001 + \cdots \\
&= (1/4) + (1/4)^2 + (1/4)^3 + \cdots \\
&= \frac{1/4}{1 - (1/4)} \\
&= 1/3.
\end{aligned}
$$

2.4.2 Theorem. *There are uncountably many subsets of the positive integers.*

Proof. If $A \subseteq \mathbb{Z}^+$, then A can be represented as an infinite sequence of 0's and 1's, by specifying that the ith term of the sequence is 1 if the positive integer i belongs to A, and 0 if i does not belong to A. Thus if $1 \in A$, $2 \in A$, $3 \notin A$, $4 \in A$, $5 \notin A$, $6 \notin A$, ..., the sequence is $110100\ldots$. This sequence corresponds to the real number $.110100$ in binary. Now an objection can be raised; the subsets $\{2, 5\}$ and $\{2, 6, 7, 8, 9, \ldots\}$ are represented by the same real number, since $.01001 = .010001111\ldots$. But this ambiguity *increases* the number of subsets relative to the reals, so the conclusion is unaffected. We know that there are uncountably many reals between 0 and 1, and therefore there are uncountably many subsets of $\mathbb{Z}^+$. ∎

How many *finite* subsets of the positive integers are there? For any fixed positive integer n, there are certainly only finitely many subsets of $\{1, \ldots, n\}$. (There are 2^n such subsets; see Section 2.1, Problem 4.) If A is the collection of all finite subsets of $\mathbb{Z}^+$, and A_n is the set of all subsets of $\{1, \ldots, n\}$, then $A = \bigcup_{n=1}^{\infty} A_n$. Thus we have countably many sets $A_1, A_2, \ldots$, and each A_n is finite.

We can conclude that A, the union of the A_n, is countable. In fact, we have a stronger result.

2.4.3 Theorem. *A countable union of countable sets is countable. In other words, if each of the sets $A_1, A_2, \ldots$ is countable, then $A = \bigcup_{n=1}^{\infty} A_n$ is countable.*

Proof. Since each A_n is countable, its elements $a_{n1}, a_{n2}, \ldots$ can be listed. Thus we can arrange the elements of A in the following array:

$$
\begin{aligned}
A_1 &: a_{11} \quad a_{12} \quad \ldots \\
A_2 &: a_{21} \quad a_{22} \quad \ldots \\
A_3 &: a_{31} \quad a_{32} \quad \ldots \\
&\vdots
\end{aligned}
$$

We can then count A by the same diagonal scheme that we used to count the rationals. ∎

Problems For Section 2.4

1. Find the flaw in the following "proof" that the rationals are uncountable. If the rationals between 0 and 1 are countable, they can be arranged in a list exactly as in the argument that the reals are uncountable:

$$
\begin{aligned}
x_1 &= .r_{11}r_{12}r_{13}\ldots \\
x_2 &= .r_{21}r_{22}r_{23}\ldots \\
x_3 &= .r_{31}r_{32}r_{33}\ldots \\
&\vdots
\end{aligned}
$$

Now pick a rational number $r = .r_1 r_2 \ldots$ such that for every n, $r_n \neq r_{nn}$. Since r does not appear on the list, it follows that the rationals are uncountable.

2. If $A_1, A_2, \ldots, A_n$ are countable sets, show that the Cartesian product $A_1 \times \cdots \times A_n$ is countable.

3. Let $r_1, r_2, r_3, \ldots$ be a list of all the positive rationals. Show that it is not possible for the list to be in increasing order ($r_1 < r_2 < r_3 < \cdots$).

4. Let A be any infinite set, countable or uncountable. Show that there is a one-to-one mapping from $\mathbb{Z}^+$ into A, so that it is reasonable to say that A is at least as large as $\mathbb{Z}^+$.

3

Elementary Number Theory

3.1 The Euclidean Algorithm

Many results in mathematics are abstractions of familiar properties of the integers, so elementary number theory will be useful background for us, in addition to being very interesting in its own right.

We start with the familiar fact that if a and b are integers with $b \neq 0$, we may divide a by b to produce a quotient q and remainder r, with $|r| < |b|$:

$$a = qb + r, \qquad |r| < |b|.$$

For example, if $a = -53$, $b = 8$, we have $-53 = (-7)8 + 3$, so $q = -7$, $r = 3$. We can also write $-53 = (-6)8 - 5$, so $q = -6$, $r = -5$. Thus q and r are not unique, but we can force uniqueness by specifying that

$$0 \leq r < |b|.$$

If $r = 0$, we say that b *divides* a, and write $b|a$. We are going to show that any two integers a and b (not both zero) have a *greatest common divisor* d, defined as follows:

(i) $d|a$ and $d|b$;
(ii) If c is any integer that divides both a and b, then $c|d$.

The greatest common divisor of a and b is denoted by $\gcd(a, b)$. (The gcd is unique up to multiplication by -1; the uniqueness question will be discussed at the end of the section.)

We'll give an algorithm, called the *Euclidean algorithm*, for finding $\gcd(a, b)$. The following result is needed to establish the validity of the algorithm.

3.1.1 Lemma. *If $a = qb + r$, $b \neq 0$, then $\gcd(a, b) = \gcd(b, r)$.*

Proof. Let $d = \gcd(b, r)$; then $d|b$ and $d|r$, hence $d|a$. If $c|a$ and $c|b$, then c divides $a - qb$, in other words, c divides r. But then c divides both b and r, hence $c|d$. ∎

Rather than give an abstract description of the algorithm, we illustrate it with a typical example, namely, $\gcd(54, 123)$. Divide the larger number $a = 123$ by the smaller number $b = 54$ to find a quotient and remainder:

$$123 = 2(54) + 15, \qquad \text{so } q = 2, \ r = 15. \tag{1}$$

(Nothing more complicated than ordinary long division is involved here.) By Lemma 3.1.1, $\gcd(123, 54)$ (which is the same as $\gcd(54, 123)$) $= \gcd(54, 15)$. Divide 54 by 15 to get

$$54 = 3(15) + 9, \tag{2}$$

so again by (3.1.1), $\gcd(54, 15) = \gcd(15, 9)$. As we continue this process, the remainders get smaller, and the algorithm terminates when a zero remainder is reached:

$$15 = 1(9) + 6, \quad \text{and} \quad \gcd(15, 9) = \gcd(9, 6) \tag{3}$$

$$9 = 1(6) + 3, \quad \text{and} \quad \gcd(9, 6) = \gcd(6, 3) \tag{4}$$

$$6 = 2(3) + 0, \quad \text{and} \quad \gcd(6, 3) = \gcd(3, 0) = 3 \tag{5}$$

(Note that since 3 is a divisor of 6, the greatest common divisor of 6 and 3 must be 3, so the "$\gcd(3, 0)$" step can be omitted if desired.)

Therefore, $\gcd(54, 123) = 3$. In general, $\gcd(a, b)$ will be the last divisor in the computation.

The following result is a corollary of the Euclidean algorithm.

3.1.2 Theorem.   *If $d = \gcd(a, b)$, then there are integers s and t such that $sa + tb = d$.*

Proof (informal).   Again we use $a = 123$, $b = 54$ as an example with all features of the general case. The five division computations needed to compute d are labeled (1)–(5) above. Start with Equation (4):

$$3 = 9 - 1(6);$$

9 appears as the remainder in (2) and 6 as the remainder in (3). Substitute to obtain

$$3 = 54 - 3(15) - 1\big(15 - 1(9)\big) = 54 - 4(15) + 1(9).$$

Now use Equation (1) to replace 15 by $123 - 2(54)$, and substitute $54 - 3(15)$ for 9 via Equation (2). The result is

$$3 = 54 - 4\big(123 - 2(54)\big) + 1\big(54 - 3(15)\big)$$
$$= -4(123) + 10(54) - 3(15)$$
$$= -4(123) + 10(54) - 3\big(123 - 2(54)\big)$$
$$= -7(123) + 16(54)$$

as desired. The point is that successive substitution will express d in terms of remainders that are higher on the list until, finally, the first equation is reached. At this point, d can be written as a linear combination of a and b. ∎

Remark. I believe that the above argument is more convincing than an abstract proof, and just as precise. In order to be sure that the abstract version is sound, I need to see how it works in a typical case. But having seen a typical case such as the one above, I then tend to find the abstract discussion—which simply substitutes Greek letters for the numbers 123, 54, etc.—unnecessary. In this situation, abstraction does not add anything significant, and definitely interferes with the learning process.

The above computation can be done more systematically using the following format, to be explained in a moment. The algorithm to be presented is quite efficient for both hand and machine computation.

i	q_i	s_i	t_i	r_i
-1		1	0	123
0		0	1	54
1	2	1	-2	15
2	3	-3	7	9
3	1	4	-9	6
4	1	-7	16	3

In the first row of the table, we enter the initial data

$$s_{-1} = 1, \quad t_{-1} = 0, \quad r_{-1} = s_{-1}a + t_{-1}b = 1(123) + 0(54) = 123.$$

In the second row, we enter

$$s_0 = 0, \quad t_0 = 1, \quad r_0 = s_0a + t_0b = 54.$$

To complete the table, we take q_i to be the quotient when r_{i-2} is divided by r_{i-1}, and

$$s_i = s_{i-2} - q_i s_{i-1}, \quad t_i = t_{i-2} - q_i t_{i-1}, \quad r_i = r_{i-2} - q_i r_{i-1}.$$

Thus the q_i and r_i, $i \geq 1$, will be the successive quotients and remainders in the Euclidean algorithm.

Let's compute all the numbers. We have $q_1 = 2$ and

$$s_1 = s_{-1} - q_1 s_0 = 1 - 2(0) = 1, \quad t_1 = t_{-1} - q_1 t_0 = 0 - 2(1) = -2,$$
$$r_1 = r_{-1} - q_1 r_0 = 123 - 2(54) = 15,$$

which is the first remainder of the computation. Similarly, $q_2 = 3$ and

$$s_2 = s_0 - q_2 s_1 = 0 - 3(1) = -3, \quad t_2 = t_0 - q_2 t_1 = 1 - 3(-2) = 7,$$
$$r_2 = r_0 - q_2 r_1 = 54 - 3(15) = 9,$$

the second remainder; $q_3 = 1$ and

$$s_3 = s_1 - q_3 s_2 = 1 - 1(-3) = 4, \quad t_3 = t_1 - q_3 s_2 = -2 - 1(7) = -9,$$
$$r_3 = r_1 - q_3 r_2 = 15 - 1(9) = 6,$$

the third remainder; $q_4 = 1$ and

$$s_4 = s_2 - q_4 s_3 = -3 - 1(4) = -7, \quad t_4 = t_2 - q_4 t_3 = 7 - 1(-9) = 16,$$
$$r_4 = r_2 - q_4 r_3 = 9 - 1(6) = 3,$$

the fourth remainder. We stop the computation at this point because r_4 divides r_3. At each stage, we have

$$s_i a + t_i b = r_i.$$

This is true for $i = -1$ and $i = 0$, by definition of r_{-1} and r_0, and it holds for $i \geq 1$ by induction. To see this, suppose that

$$s_{i-2} a + t_{i-2} b = r_{i-2} \qquad \text{and} \qquad s_{i-1} a + t_{i-1} b = r_{i-1}.$$

Multiply the second equation by $-q_i$ and add it to the first equation; the result is $s_i a + t_i b = r_i$, as desired. For our numerical example, we have

$$r_4 = s_4 a + t_4 b = -7(123) + 16(54) = 3,$$

which agrees with the result obtained by successive substitution.

The greatest common divisor is not unique, because if d satisfies the defining conditions (i) and (ii), so does $-d$. It is convenient to choose the positive gcd and call it *the* greatest common divisor. Uniqueness must still be proved, and this will make a good exercise (Problem 1).

Problems For Section 3.1

1. Let both d and e be greatest common divisors of a and b. Show that $|d| = |e|$.

2. If $d = \gcd(a, b)$, show that $|d|$ is, as the terminology suggests, the greatest integer that divides both a and b.

3. If $a = 770$, $b = 84$, find the greatest common divisor of a and b, and express the gcd as a linear combination of a and b.

4. Repeat Problem 3 for $a = 232$, $b = 14$.

5. If $d = \gcd(a, b)$, are the integers s and t such that $sa + tb = d$ unique? Prove or give an explicit counterexample. (This question will be considered in detail in Section 3.5.)

3.2 Unique Factorization

One of the most important consequences of the Euclidean algorithm is that every integer (other than 0 and ± 1) has a unique representation as a product of primes. We proceed directly to this.

3.2.1 Definition. Let a be an integer, not 0 or 1 or -1. We say that a is *composite* if it can be expressed in the form $a = bc$, where $|b| > 1$ and $|c| > 1$. If a is not composite, it is said to be *prime*. Equivalently, a is prime if and only if the only divisors of a are ± 1 and $\pm a$. Note that if p is prime, so is $-p$. Thus if we know all primes between 2 and n, we automatically know all primes between $-n$ and n. (For this reason, many authors use "prime" and "composite" to refer only to positive integers greater than 1.)

The first few positive primes are $2, 3, 5, 7, 11, 13, 17, 19, 23, 29, 31, 37, 41, \ldots$. The following result was discovered by Euclid.

3.2.2 Theorem. *If p is prime and p divides a product ab, then either p divides a or p divides b.*

Proof. Suppose that p does not divide a, and let $d = \gcd(a, p)$. Since $d|p$ and p is prime, d must be p or 1. If $d = p$ then $p|a$, contradicting our assumption. Consequently, $d = 1$, and by (3.1.2), there are integers s and t such that $sa + tp = 1$. Multiply by b to obtain $sab + tpb = b$. Since $p|ab$, it follows that p divides $sab + tpb$, and therefore $p|b$. ∎

3.2.3 Corollary. *If p is prime and p divides a product $a_1 a_2 \cdots a_n$, then p divides at least one of the a_i.*

Proof. Mathematical induction works very well here. The case $n = 2$ has just been proved. If the result holds for all products of n integers, it must hold for all products $a_1 \cdots a_{n+1}$. To see this, write $a_1 \cdots a_{n+1} = (a_1 \cdots a_n)a_{n+1}$ and apply the $n = 2$ case and the induction hypothesis. ∎

The *existence* of a factorization of an arbitrary integer a (not 0 or ± 1) into a product of primes is easy to establish. If a is prime, we are finished. If a is composite, we can write $a = bc$ with $|b| > 1$ and $|c| > 1$, hence $|b| < |a|$ and $|c| < |a|$. If b and c are prime, we are finished. If b is composite, then $b = de$ with $|d| < |b|$, $|e| < |b|$, and so on. The process must terminate in a finite number of steps, and upon completion we have a prime factorization. (Formally, we are using the fact that any nonempty subset of the nonnegative integers has a smallest element. This idea will be discussed in detail in Section 4.1.)

To prove *uniqueness* of factorization, we must use the theory we have developed in this chapter.

3.2.4 Unique Factorization Theorem. *Let a be an integer, not 0 or 1 or -1. Then*
(1) *a can be written as a product $p_1 p_2 \cdots p_n$, where each p_i is prime.*
(2) *If $a = p_1 \cdots p_n = q_1 \cdots q_m$, where all p_i and q_j are prime, then $n = m$ and after suitable renumbering, $p_i = \pm q_i$ for all $i = 1, 2, \ldots, n$.*

Proof. Part (1) follows from the discussion preceding the statement of the theorem. To prove (2), note that p_1 divides $p_1 \cdots p_n$, and hence p_1 divides $q_1 \cdots q_m$. By (3.2.3), p_1 divides one of the q_i. Since p_1 and q_i are prime, we have $p_1 = \pm q_i$. Renumber so that $i = 1$; then $p_1 p_2 \cdots p_n = (\pm p_1)q_2 \cdots q_m$, hence $p_2 \cdots p_n = \pm q_2 \cdots q_m$. Now p_2 divides $p_2 \cdots p_n$, and therefore p_2 divides $\pm q_2 \cdots q_m$. The above argument may be repeated to obtain, after renumbering, $p_2 = \pm q_2$. Continuing to match p's and q's in this fashion, we conclude that $m = n$ and $p_i = \pm q_i$ for all i. (If, for example, $n = 3$ and $m = 5$, we would have $p_1 p_2 p_3 = \pm p_1 p_2 p_3 q_4 q_5$, so that $q_4 q_5 = \pm 1$, which is impossible since $|q_4|$ and $|q_5|$ are at least 2.) ∎

If $p_1, p_2, \ldots$ are the positive primes in increasing order, and a is an arbitrary integer ≥ 2, the prime factorization of a may be written in the convenient form

$$a = p_1^{e_1} p_2^{e_2} \cdots p_k^{e_k}.$$

By the Unique Factorization Theorem, the integer k and the exponents e_i are uniquely determined by a.

3.2.5 Example.   Find the prime factorization of 69972.

An inefficient but straightforward procedure is to divide successively by powers of $p_1, p_2, \ldots$. Thus,

$$\frac{69972}{2} = 34986; \quad \frac{34986}{2} = 17493,$$

which is not divisible by 2. Now $\frac{17493}{3} = 5831$, which is not divisible by 3 or 5. Continuing,

$$\frac{5831}{7} = 833; \quad \frac{833}{7} = 119; \quad \frac{119}{7} = 17,$$

which is prime. The result is

$$69972 = (2^2)(3)(7^3)(17)$$

The greatest common divisor of two integers a and b can be expressed in terms of the prime factorizations of a and b, as follows.

3.2.6 Theorem.   *Suppose that the prime factorizations of a and b are*

$$a = p_1^{e_1} p_2^{e_2} \cdots p_k^{e_k}, \quad b = p_1^{f_1} p_2^{f_2} \cdots p_k^{f_k}$$

(where e_i, $f_i \geq 0$; if $e_i = 0$, then p_i does not appear in the factorization of a). Then the greatest common divisor of a and b is

$$d = p_1^{g_1} p_2^{g_2} \cdots p_k^{g_k} \quad \text{where } g_i = \min(e_i, f_i).$$

Proof.   Since $g_i \leq e_i$ and $g_i \leq f_i$, d divides both a and b. By the Unique Factorization Theorem, any divisor c of a and b must have the form $p_1^{h_1} p_2^{h_2} \cdots p_k^{h_k}$, where $h_i \leq e_i$ and $h_i \leq f_i$. Consequently, $h_i \leq g_i$, so c divides d. ∎

Problems For Section 3.2

1.   Find the prime factorization of 10561485.

2.   Prove (following Euclid) that there are infinitely many primes. [If $p_1, \ldots, p_k$ are the only positive primes, let $N = p_1 p_2 \cdots p_k + 1$, and derive a contradiction.]

3.   Show that there are arbitrarily large gaps in the sequence of primes, that is, for every positive integer n there is a positive integer N such that $N + 1, \ldots, N + n$ are all composite. [Take $N = t(n + 1)! + 1$ for any positive integer t.]

4.   (The sieve of Eratosthenes) Suppose that we wish to list all primes $p \leq n$. Start with all positive integers from 1 to n, and cross out strict multiples of $p_1 = 2$, i.e., remove $4, 6, 8, \ldots$ from the list. Then cross out all strict multiples of $p_2 = 3$, and repeat for $p_3 = 5$, $p_4 = 7$, $p_5 = 11$, and so on. Show that after all primes $p \leq \sqrt{n}$ have been considered and their strict multiples removed from the list, we are left with the set of primes between 1 and n.

5. If n and k are positive integers but $\sqrt[k]{n}$ is not an integer, show that $\sqrt[k]{n}$ is irrational. [If $\sqrt[k]{n} = a/b$ where a and b are integers, then $a^k = b^k n$. Use the Unique Factorization Theorem to produce a contradiction.]

6. The *least common multiple* (lcm) of a and b is an integer m such that
  (i) $a|m$ and $b|m$, and
 (ii) if $a|c$ and $b|c$ then $m|c$.
      (a) If the prime factorizations of a and b are

$$a = p_1^{e_1} \cdots p_k^{e_k}, \quad b = p_1^{f_1} \cdots p_k^{f_k},$$

      find $\mathrm{lcm}(a, b)$ in terms of the p_i.
      (b) Show that $ab = \gcd(a, b) \, \mathrm{lcm}(a, b)$.

7. If $m = \mathrm{lcm}(a, b)$, show that $|m|$ is the smallest positive integer that is a multiple of a and b.

8. If $a = (2^3)(5^2)(7)(13)$ and $b = (2^2)(5)(13^2)$, find $\gcd(a, b)$ and $\mathrm{lcm}(a, b)$.

3.3 Algebraic Structures

The integers are so familiar that we work with them instinctively. For example, if in a computation we need to replace $36 + 45$ with $45 + 36$, we just do it without fear of being challenged. But suppose we ask if there are other mathematical systems in which the basic properties of the integers remain valid. To attempt to answer this question, we are forced to think about what these properties are. First of all, addition and multiplication are examples of *binary operations*. In other words, given any two integers a and b, we can operate on them to produce $a + b$ and ab. If R is any nonempty set, a binary operation on R is simply a function f on $R \times R$. Thus given $a, b \in R$, the operation produces an element $f(a, b)$. If R is the set $\mathbb{Z}$ of integers and f is addition or multiplication, then $f(a, b)$ belongs to R, but this need not hold in general. For example, if a and b are integers and $b \neq 0$, then a/b is not necessarily an integer.

Suppose that R is a nonempty set, and two binary operations are defined on R. We will write the operations as addition and multiplication, but keep in mind that R need not be $\mathbb{Z}$, so at this point we do not have a clear picture of what $a + b$ and ab look like. Here is a list of highly desirable properties; the integers satisfy all except the last.

(A1)	*Closure under addition*:	If a and b belong to R, then $a + b$ is also in R.
(A2)	*Associativity of addition*:	$a + (b + c) = (a + b) + c$ for all a, b, c in R.
(A3)	*Additive identity*:	There is an element 0 in R such that $a + 0 = 0 + a = a$ for all a in R.
(A4)	*Additive inverse*:	For each a in R there is an element $-a$ in R such that $a + (-a) = (-a) + a = 0$.
(A5)	*Commutativity of addition*:	$a + b = b + a$ for all a, b in R.

(M1) *Closure under multiplication*:   If a and b belong to R, then ab is also in R.

(M2) *Associativity of multiplication*: $a(bc) = (ab)c$ for all a, b, c in R.

(M3) *Multiplicative identity*:   There is an element 1 in R such that $a1 = 1a = a$ for all a in R.

(M4) *Distributive laws*: $a(b + c) = ab + ac$ and $(a + b)c = ac + bc$ for all a, b, c in R.

(M5) *Commutativity of multiplication*: $ab = ba$ for all a, b in R.

(M6) *No zero-divisors*:   If a and b belong to R and $ab = 0$, then either $a = 0$ or $b = 0$.

(M7) *Multiplicative inverse*:   If a belongs to R and $a \neq 0$, there is an element a^{-1} in R such that $aa^{-1} = a^{-1}a = 1$.

An *algebraic structure* is specified by selecting properties from this list; the properties selected are called the *axioms* of the particular structure.

If R satisfies (A1)–(A4), we say that R forms a *group* under addition; if R also satisfies (A5), then R is an *abelian group* (or *commutative group*).

If R satisfies (A1)–(A5) and (M1)–(M4), R is said to be a *ring* under the two given binary operations. If (M5) holds as well, we have a *commutative ring*. If we tack on (M6), we get an *integral domain*. Finally, if we add (M7) so that all properties on the list are satisfied, we say that R is a *field*.

Intuitively, a ring is a set in which we can do addition, subtraction ($a-b = a+(-b)$), and multiplication without leaving the set. An integral domain is a commutative ring in which cancellation is possible; if $ab = ac$ and $a \neq 0$, then, since $a(b - c) = 0$, we must have $b = c$. A field is a set in which we can do addition, subtraction, multiplication and division (by a nonzero denominator) without leaving the set.

The integers are an integral domain. Familiar examples of fields are the rational numbers, the real numbers and the complex numbers. We are now going to concentrate on a ring that arises in a natural way from elementary number theory.

3.3.1 Definitions and Comments. If a and b are integers and m is a positive integer ≥ 2, we say that a *is congruent to* b modulo m, written

$$a \equiv b \bmod m,$$

if $a - b$ is divisible by m.

Congruence modulo m is an equivalence relation which partitions the integers into m equivalence classes $S(i)$, $0 \leq i \leq m - 1$. Explicitly,

$$S(i) = \{\ldots, -2m + i, \ -m + i, \ i, \ m + i, \ 2m + i, \ldots\},$$

the set of integers that leave remainder i when divided by m. In (1.6.3) we verified this for the case $m = 4$; the general case is identical, with 4 replaced by m.

The sets $S(i)$ are called *residue classes* modulo m. If x belongs to $S(i)$, then i is called the *residue* of x modulo m; i is simply the remainder when x is divided by m.

Let's again examine the case $m = 4$. Suppose that the number $5 \in S(1)$ is added to the number $11 \in S(3)$. The result is $16 \in S(0)$. In fact, if 5 is replaced by an arbitrary element of $S(1)$, and 11 by an arbitrary element of $S(3)$, the sum will still belong to $S(0)$. For example, $13 + (-21) = -8 \in S(0)$. A similar result holds for multiplication. We have $5(11) = 55 \in S(3)$, and the product of any number in $S(1)$ by any number in $S(3)$ belongs to $S(3)$. The results can be generalized, as follows.

3.3.2 Theorem.   *If $x \equiv y \bmod m$ and $x' \equiv y' \bmod m$, then*

(a) $x + x' \equiv y + y' \bmod m$ (*similarly, $x - x' \equiv y - y' \bmod m$*)

(b) $xx' \equiv yy' \bmod m$.

Proof.   We have $x + x' - (y + y') = (x - y) + (x' - y')$. By hypothesis, m divides $x - y$ and $x' - y'$, and therefore m divides $(x - y) + (x' - y') = x + x' - (y + y')$, proving (a). To prove (b), write

$$xx' - yy' = xx' - xy' + xy' - yy' = x(x' - y') + (x - y)y'.$$

(This is a standard trick; we subtract and add back the same quantity, and then factor the resulting expression.) By hypothesis, m divides $x' - y'$ and $x - y$, and consequently, m divides $xx' - yy'$. ∎

Thus if we add, subtract, or multiply, and we are interested only in the residue modulo m of the result, it makes no difference which member of a particular residue class is used in the computation. Frequently, this property is expressed by saying that the equivalence relation $\equiv$ is *compatible* with addition and multiplication, or that $\equiv$ is a *congruence relation* with respect to addition and multiplication. The basic point is that we can define addition and multiplication of residue classes in a natural way. If $[x]$ is the residue class of x, define

$$[x] + [x'] = [x + x']$$

$$[x][x'] = [xx']$$

By (3.3.2), the definitions are sensible. For if $[x] = [y]$ and $[x'] = [y']$ then $x \equiv y \bmod m$ and $x' \equiv y' \bmod m$, hence $x + x' \equiv y + y' \bmod m$ and $xx' \equiv yy' \bmod m$. Thus $[x + x'] = [y + y']$ and $[xx'] = [yy']$.

Attention. Always check that an object that you are defining is in fact *well-defined*. This is a reflex that an aspiring mathematician develops quickly. In the present case we are defining addition and multiplication of residue classes, but when we write $[x]$, the residue class of x, we are explicitly choosing one particular member of the class, namely x. (If x belongs to the residue class S, then x is said to be a *representative* of S.) What if we were to choose another representative, say y? Then $[x] = [y]$, and similarly if y' belongs to the residue class $[x']$, then $[x'] = [y']$. *We must make sure that addition and multiplication of residue classes does not depend on the particular representatives that we select.* In other words, we must prove that $[x + x'] = [y + y']$ and $[xx'] = [yy']$, as we have just done.

We may choose $0, 1, \dots, m - 1$ as a *canonical system of representatives*. In other words, from the residue class $S(i)$ we choose i, $0 \le i \le m - 1$. Then the residue classes can be written as $[0], [1], \ldots, [m-1]$. Addition and multiplication tables are given below for the case $m = 4$. As is customary, we denote the residue class $[x]$ simply by x; the set $\mathbb{Z}_m$ of residue classes mod m is then written as $\{0, 1, \ldots, m-1\}$.

+	0	1	2	3
0	0	1	2	3
1	1	2	3	0
2	2	3	0	1
3	3	0	1	2

×	0	1	2	3
0	0	0	0	0
1	0	1	2	3
2	0	2	0	2
3	0	3	2	1

The set $\mathbb{Z}_m$ forms a commutative ring under the operations of addition and multiplication; $\mathbb{Z}_m$ is often referred to as the *ring of integers modulo* m. To check that $\mathbb{Z}_m$ is a commutative ring, we must verify that $\mathbb{Z}_m$ satisfies axioms (A1)–(A5) and (M1)–(M5). The computations are routine and are based on the fact that the integers form a commutative ring. To illustrate the pattern, let's verify the distributive law

$$[a]([b] + [c]) = [a][b] + [a][c].$$

This is equivalent to $[a]([b+c]) = [ab] + [ac]$, that is, $[a(b+c)] = [ab+ac]$. Thus we must verify that $a(b+c) \equiv ab + ac \bmod m$. But

$$a(b+c) - (ab + ac) = 0,$$

hence

$$a(b+c) - (ab + ac) \equiv 0 \bmod m,$$

and the result follows.

If m is composite, so that $m = ab$ where $a > 1$ and $b > 1$, then $\mathbb{Z}_m$ is not an integral domain. For $[a][b] = [m] = [0]$, so that $[a]$ and $[b]$ are zero-divisors.

Furthermore, if m is composite then $\mathbb{Z}_m$ is not a field. For if $[a]$ has a multiplicative inverse $[c]$, so that $[a][c] = [1]$, multiply both sides by $[b]$ to get $[m][c] = [b]$. But $[m] = [0]$, and we conclude that $[0] = [b]$, a contradiction.

In fact any field F is an integral domain, so that once we know that $\mathbb{Z}_m$ is not an integral domain, it cannot possibly be a field; the above computation is unnecessary. For if a and b belong to F and $ab = 0$ with $a \neq 0$, multiply both sides by a^{-1} to get $b = a^{-1}0 = 0$. (We must be careful here; we are implicitly using the fact that if c is an element in an *arbitrary* ring R, then $c0 = 0$. To see this, note that if a is any element of R, we have $c(a + 0) = ca + c0$, i.e., $ca = ca + c0$. We may subtract ca from both sides to get $0 = c0$.)

In the next section, we will show that if m is prime, then $\mathbb{Z}_m$ is a field.

Problems For Section 3.3

1. In $\mathbb{Z}_5$, compute $3(0)$, $3(1)$, $3(2)$, $3(3)$, and $3(4)$, and from your computation, find the multiplicative inverse of 3.

2. As in Problem 1, find the multiplicative inverse of each nonzero element in $\mathbb{Z}_5$.

3. Let F be a field, and let $F[X]$ be the set of all polynomials with coefficients in F. Thus a typical element of $F[X]$ is of the form

$$f(X) = a_n X^n + \cdots + a_1 X + a_0 \qquad (n = 0, 1, 2, \ldots)$$

with all $a_i \in F$. If $f(X)$ is not the zero polynomial, then the leading coefficient a_n is not zero. (Of course, we can write $X^2 + 3X - 7$ as $0X^3 + X^2 + 3X - 7$, but let's agree not to do this.)

Consider $F[X]$ with ordinary addition and multiplication of polynomials. Is $F[X]$ a commutative ring? An integral domain? A field?

4. The integers form an integral domain but not a field. However, if we consider quotients a/b where a and b are integers and $b \neq 0$, we obtain the field of rational numbers. Suppose that we pursue this idea with the integers replaced by $F[X]$ (see Problem 3). What field is obtained?

3.4 Further Properties of Congruence Modulo m

In Section 3.3 we showed that if m is composite, then $\mathbb{Z}_m$, the ring of integers modulo m, is not an integral domain. We now prove that if p is prime, then $\mathbb{Z}_p$ is not only an integral domain, but in fact a field.

3.4.1 Theorem. *If p is prime and a is not congruent to 0 mod p, then for some integer b we have $ab \equiv 1$ mod p. Thus $[a][b] = [1]$ in $\mathbb{Z}_p$, so that $[b]$ is the multiplicative inverse of $[a]$ in $\mathbb{Z}_p$. Consequently, $\mathbb{Z}_p$ is a field.*

Proof. Since p is prime and p does not divide a, the only possible common factors of a and p are ± 1. Thus the greatest common divisor of a and p is 1, and it follows from (3.1.2) that there are integers s and t such that $sa + tp = 1$. Thus $sa + tp \equiv 1$ mod p. But $tp \equiv 0$ mod p (since p is a divisor of tp). Therefore $sa \equiv 1$ mod p, and we may take $b = s$. ∎

3.4.2 Example. Find the inverse of 5 mod 17.

The proof of (3.4.1) provides a method. Using the Euclidean algorithm, we express $\gcd(5, 17)$ $(= 1)$ as a linear combination of 5 and 17. The computation is summarized in the table below.

i	q_i	s_i	t_i	r_i
-1		1	0	17
0		0	1	5
1	3	1	-3	2
2	2	-2	7	1

We have $-2(17) + 7(5) = 1$, hence $7(5) \equiv 1$ mod 17. Therefore $5^{-1} = 7$ in $\mathbb{Z}_{17}$.

Since 17 is not a large number, it is feasible in this case to compute all products $5x$, $x = 1, 2, \ldots, 16$, until we reach $5x \equiv 1$ mod 17. However, this method becomes unwieldy for large p.

Suppose that by a different computation, we found that $5(-27) \equiv 1 \bmod 17$, and concluded that the inverse of 5 is $-27 \bmod 17$. The result is correct, because -27 and 7 belong to the same residue class mod 17; 7 is the canonical representative.

The technique used to prove (3.4.1) yields an additional result, which we prove after some old and some new terminology.

3.4.3 Definitions. Recall from (2.3.3) that the integers a and m are *relatively prime* (we also say that a is *relatively prime to* m) if the only common factors of a and m are ± 1. Equivalently, $\gcd(a, m) = 1$. If a has a multiplicative inverse mod m, in other words, if there is an integer b such that $ab \equiv 1 \bmod m$, then a (and similarly b) is said to be a *unit* mod m.

3.4.4 Theorem. *The integer a is relatively prime to m if and only if a is a unit* mod m.

Proof. Suppose that a is relatively prime to m. Let s and t be integers such that $sa + tm = \gcd(a, m) = 1$. Since $tm \equiv 0 \bmod m$, we have $sa \equiv 1 \bmod m$, so a is a unit mod m. Conversely, if $ab \equiv 1 \bmod m$, then $ab - 1 \equiv 0 \bmod m$, hence $ab - 1 = mx$ for some integer x. If $d = \gcd(a, m)$ then $d|ab$ and $d|mx$, so $d|1$, i.e., $d = 1$. Thus a is relatively prime to m. ■

Further facts about relative primeness will be needed later.

3.4.5 Theorem.
(a) If c divides ab and a and c are relatively prime, then c divides b.
(b) If a and b are relatively prime to m, then ab is relatively prime to m.
(c) If $ax \equiv ay \bmod m$ and a is relatively prime to m, then $x \equiv y \bmod m$.
(d) If $d = \gcd(a, b)$, then $\frac{a}{d}$ and $\frac{b}{d}$ are relatively prime.
(e) If $ax \equiv ay \bmod m$ and $d = \gcd(a, m)$, then $x \equiv y \bmod m/d$.
(f) If $a_1, \ldots, a_k$ each divide b, and a_i and a_j are relatively prime whenever $i \neq j$, then the product $a_1 \cdots a_k$ divides b.

Proof.
(a) If p^e occurs in the prime factorization of c, and therefore in the prime factorization of ab, then it must also occur in the prime factorization of b. (Otherwise p will be a factor of a, contradicting the assumption that a and c are relatively prime.) It follows that c divides b.
(b) If p is a prime factor of ab, then either p divides a or p divides b, by (3.2.2). Since a and b are relatively prime to m, p cannot divide m.
(c) If $ax \equiv ay \bmod m$, then m divides $a(x - y)$. By (a), m divides $x - y$, so $x \equiv y \bmod m$.
(d) If c is a divisor of both a/d and b/d, then cd divides both a and b. (For if $a/d = jc$ and $b/d = kc$, then $a = jcd$ and $b = kcd$.) But then cd divides d, so that $c = \pm 1$.
(e) As in (c), m divides $a(x - y)$, hence m/d divides $(a/d)(x - y)$. But by (d), m/d and a/d are relatively prime, so by (a), m/d divides $x - y$.
(f) If p^e appears in the prime factorization of $a_1 \cdots a_k$, then since a_i and a_j are relatively prime for $i \neq j$, p^e must occur in some a_i. Since a_i divides b, it follows that p^e

divides b. Repeat for all prime factors of $a_1 \cdots a_k$ to conclude that $a_1 \cdots a_k$ divides b. ∎

Problems For Section 3.4

1. Find the inverse of $21 \bmod 37$.

2. Find the inverse of $10 \bmod 127$, and solve the equation $10x \equiv 7 \bmod 127$.

In Problems 3, 4, and 5, N is a positive integer whose ith decimal digit is a_i, that is,

$$N = a_1 + a_2 10^1 + \cdots + a_{n-1}10^{n-2} + a_n 10^{n-1}.$$

For example, if $N = 475$, then $a_1 = 5$, $a_2 = 7$, $a_3 = 4$, and $N = 5 + 7(10) + 4(10^2)$.

3. Show that N is congruent modulo 9 to the sum of its decimal digits. For example, $475 \equiv 4 + 7 + 5 \equiv 16 \equiv 1 + 6 \equiv 7 \bmod 9$. This is the basis for the familiar procedure of "casting out 9's" when checking computations in arithmetic.

4. Show that N is congruent modulo 11 to $a_1 - a_2 + a_3 - a_4 + \cdots$. For example, $6384 \equiv 4 - 8 + 3 - 6 = -7 \equiv 4 \bmod 11$.

5. Let $M = 2^r$, $r = 1, 2, \ldots, n$. Show that M divides N if and only if M divides the number consisting of the first r digits of N, i.e.,

$$M|(a_1 + a_2 10^1 + \cdots + a_r 10^{r-1}).$$

For example (with $r = 3$), we have $8|536$, and it follows that $8|29536$.

6. Prove that there are infinitely many primes of the form $4n+3$, n a positive integer. [If $p_1, \ldots, p_k$ are the only such primes, let $N = 4p_1 \cdots p_k - 1$. Since $N \equiv -1 \equiv 3 \bmod 4$ and N is larger than any of the p_i, N must be composite. Show that all prime factors of N must be congruent to 1 mod 4, and hence N must be of this form also, a contradiction.]

3.5 Linear Diophantine Equations and Simultaneous Congruences

We are going to apply the theory of congruences to find all integer solutions (x, y) of the equation

$$ax + by = c, \tag{1}$$

where a, b and c are given integers; (1) is called a *linear diophantine equation*. Rather than give an abstract discussion, we'll consider a specific example,

$$123x + 54y = 12. \tag{2}$$

The first step is to find $d = \gcd(a, b)$. For $a = 123$, $b = 54$, we computed $d = 3$ in Section 3.1, and furthermore we found integers $s = -7$ and $t = 16$ such that $sa + tb = d$:

$$123(-7) + 54(16) = 3. \tag{3}$$

Multiply (3) by 4 to obtain

$$123(-28) + 54(64) = 12. \tag{4}$$

In general, if d does not divide c, then (1) has no solution. (If (1) holds, then since d divides both a and b, it must divide c.) Now (2) can be written in the form

$$123x \equiv 12 \bmod 54, \tag{5}$$

and by (4), $x = -28$ is a solution. Therefore, any number in the residue class $[-28] = [26] \bmod 54$ is also a solution. To find additional solutions, note that (5) is equivalent to the statement that 54 divides $123x - 12$, that is, $3(18)$ divides $3(41x - 4)$. It follows that 18 divides $41x - 4$, in other words,

$$41x \equiv 4 \bmod 18. \tag{6}$$

Since $x = 26$ is a solution and $26 \equiv 8 \bmod 18$, any member of the residue class $[8] \bmod 18$ is also a solution. Thus, we have found solutions of the form

$$x = 8 + 18u, \quad u \in \mathbb{Z}. \tag{7}$$

In fact there are no other solutions. For if x_1 and x_2 each satisfy (6), then

$$41(x_1 - x_2) = 41x_1 - 41x_2 \equiv 4 - 4 \equiv 0 \bmod 18,$$

and consequently 18 divides $41(x_1 - x_2)$. The key observation is that 18 and 41 are relatively prime, by (3.4.5 (d)). By (3.4.5 (a)), 18 must divide $x_1 - x_2$, hence $x_1 - x_2 = 18v$ for some $v \in \mathbb{Z}$. If $x_2 = 8 + 18u$, then $x_1 = 8 + 18(u + v)$, which is again of the form (7).

If x is given by (7), the value of y corresponding to x can be found from equation (2):

$$54y = 12 - 123x$$
$$18y = 4 - 41x = 4 - 41(8 + 18u) = -324 - 41(18u)$$
$$y = -18 - 41u. \tag{8}$$

Our example has all the features of the general case. The method may be summarized as follows.

1. Compute $d = \gcd(a, b)$ and find integers s and t such that $sa + tb = d$.

2. If d does not divide c, then (1) has no solution. If $d|c$, then

$$x = \frac{c}{d}s, \quad y = \frac{c}{d}t$$

is a solution.

3. Write (1) as $ax \equiv c \bmod b$, which by (3.4.5(e)) is equivalent to

$$\frac{a}{d}x \equiv \frac{c}{d} \bmod \frac{b}{d}.$$

[It is equally valid to write (1) as $by \equiv c \bmod a$. Since computations modulo m will generally be easier for smaller m, the form $ax \equiv c \bmod b$ will usually be preferred if $a > b$.]

If x_0 is any representative of the residue class $[\frac{c}{d}s]$ mod $\frac{b}{d}$, then x_0 is a solution. Furthermore, since $\frac{a}{d}$ and $\frac{b}{d}$ are relatively prime, all solutions must be of the form

$$x = x_0 + \frac{b}{d}u, \quad u \in \mathbb{Z}.$$

Note that although we have found infinitely many solutions, *there are only d distinct solutions* mod b. For if $x_1 = x_0 + \frac{b}{d}u$ and $x_2 = x_0 + \frac{b}{d}(u + k)$, then $x_1 - x_2 = \frac{b}{d}k$. The integers $\frac{b}{d}k$ are distinct mod b for $k = 0, 1, \ldots, d - 1$, but when $k = d$, we have $x_1 - x_2 = b \equiv 0$ mod b. For our numerical example, set $u = 0, 1, 2$ in (7) to get 3 distinct solutions mod 54: $x = 8, 26, 44$.

4. Obtain y from (1):

$$by = c - ax = c - ax_0 - \frac{ab}{d}u$$

$$y = \frac{c - ax_0}{b} - \frac{a}{d}u, \quad u \in \mathbb{Z}.$$

(To simplify the computation, (1) may be divided by d before solving for y, as we did in deriving (8).)

All of our analysis applies to an equation of the form $ax \equiv c$ mod b, because this equation has the equivalent form $ax + by = c$ (see step 3).

We now consider simultaneous congruences of the form

$$a_1 x \equiv b_1 \text{ mod } m_1$$

$$a_2 x \equiv b_2 \text{ mod } m_2$$

$$\vdots$$

$$a_k x \equiv b_k \text{ mod } m_k.$$

It is convenient to assume that $a_i = 1$ for all i, and in fact this involves no loss of generality. For the equation $ax \equiv b$ mod m, where $\gcd(a, m) = d$, is equivalent to

$$\frac{a}{d}x \equiv \frac{b}{d} \text{ mod } \frac{m}{d}. \tag{9}$$

Since $\frac{a}{d}$ and $\frac{m}{d}$ are relatively prime by (3.4.5 (d)), $\frac{a}{d}$ has a multiplicative inverse c mod $\frac{m}{d}$, by (3.4.4). Multiply (9) by c to obtain

$$x \equiv c\frac{b}{d} \text{ mod } \frac{m}{d}. \tag{10}$$

Thus, the standard form for simultaneous congruences is

$$x \equiv b_1 \text{ mod } m_1$$

$$x \equiv b_2 \text{ mod } m_2$$

$$\vdots \tag{11}$$

$$x \equiv b_k \text{ mod } m_k.$$

Without additional assumptions on the m_i, the system (11) need not have a solution. (An example: $x \equiv 2$ mod 4, $x \equiv 1$ mod 2; the first equation forces x to be even, and the second forces x to be odd.) We are going to assume that the m_i are *relatively prime in pairs,* that is,

$$\gcd(m_i, m_j) = 1 \quad \text{for } i \neq j.$$

In this case, we will find all solutions of (11).

3.5.1 Chinese Remainder Theorem. If $m_1, \ldots, m_k$ are relatively prime in pairs, then the system $x \equiv b_j \bmod m_j$, $1 \leq j \leq k$, has a solution. In fact, the set of solutions forms a single residue class modulo $m_1 m_2 \cdots m_k$, so that the system has a unique solution mod $m_1 m_2 \cdots m_k$.

Proof. Let $m = m_1 m_2 \cdots m_k$, and define

$$y_1 = \frac{m}{m_1}, \; y_2 = \frac{m}{m_2}, \ldots, y_k = \frac{m}{m_k};$$

thus y_i is the product of all the m_j's except m_i. Since the m_i are relatively prime in pairs, $\gcd(y_i, m_i) = 1$ for all i. Let z_i be an inverse of y_i mod m_i; z_i exists by (3.4.4). If $i \neq j$, then m_j divides y_i, hence $y_i \equiv 0 \bmod m_j$, and consequently $y_i z_i \equiv 0 \bmod m_j$. Thus

$$y_i z_i \equiv \begin{cases} 1 \bmod m_i \\ 0 \bmod m_j, \quad j \neq i \end{cases} \tag{12}$$

Let

$$x_0 = b_1 y_1 z_1 + b_2 y_2 z_2 + \cdots + b_k y_k z_k; \tag{13}$$

we claim that the set of solutions coincides with the residue class of x_0 mod m. For if $x \equiv x_0 \bmod m$, then m divides $x - x_0$, so m_j divides $x - x_0$ for all j. Thus $x \equiv x_0 \bmod m_j$, $1 \leq j \leq k$. By (13),

$$x \equiv \sum_{i=1}^{k} b_i y_i z_i \bmod m_j,$$

and by (12), $x \equiv b_j \bmod m_j$, $j = 1, \ldots, k$. Conversely, let x be a solution, so that $x \equiv b_j \bmod m_j$, $j = 1, \ldots, k$. By (12) and (13), $x_0 \equiv b_j \bmod m_j$, hence $x \equiv x_0 \bmod m_j$. Thus each m_j divides $x - x_0$, and since the m_j are relatively prime in pairs, the product of the m_j (namely m), divides $x - x_0$ by 3.4.5(f). In other words, $x \equiv x_0 \bmod m$. $\blacksquare$

3.5.2 Example. We will solve $x \equiv 1 \bmod 2$, $x \equiv 4 \bmod 5$, $x \equiv 4 \bmod 7$.

We have $m_1 = 2$, $m_2 = 5$, $m_3 = 7$, $m = 2(5)(7) = 70$, $y_1 = m/m_1 = 35$, $y_2 = m/m_2 = 14$, $y_3 = m/m_3 = 10$. Since $y_1 \equiv 1 \bmod 2$, $y_2 \equiv 4 \bmod 5$, $y_3 \equiv 3 \bmod 7$, we may take $z_1 = 1$, $z_2 = 4$, $z_3 = 5$. (In this case, the numbers are small enough so that z_i may be found simply by trying $1, 2, \ldots, m_i - 1$ in turn. In more complicated cases, the technique of (3.4.2) may be needed.) We have $b_1 = 1$, $b_2 = 4$, $b_3 = 4$, so by (13), one solution is given by

$$x_0 = 1(35)(1) + 4(14)(4) + 4(10)(5)$$
$$= 1(35) + 4(56) + 4(50)$$
$$= 459 \equiv 39 \bmod 70.$$

The set of solutions is described by

$$x = 39 + 70u, \quad u \in \mathbb{Z}.$$

The Chinese Remainder Theorem sets up a one-to-one correspondence between the k-tuple $(b_1, b_2, \ldots, b_k)$, where b_i is reduced modulo m_i (i.e., replaced by its canonical representative), and the number $\sum_{i=1}^{k} b_i y_i z_i$, reduced modulo m (see Problem 5). Addition modulo m can be accomplished by doing addition modulo m_i on the ith component for $i = 1, \ldots, k$. In the above example, $(1, 4, 4)$ corresponds to 39 mod 70. Similarly, $(1, 3, 6)$ corresponds to $1(35) + 3(56) + 6(50) = 503 \equiv 13 \bmod 70$. Adding these triples componentwise, we obtain

$$(1 + 1 \bmod 2, 4 + 3 \bmod 5, 4 + 6 \bmod 7) = (0, 2, 3),$$

which corresponds to $0(35) + 2(56) + 3(50) = 262 \equiv 52 = 39 + 13 \bmod 70$.

If the m_i are much smaller than m, it may be easier to do parallel additions modulo the m_i, rather than a single addition modulo m. Also, if x and y are fixed integers, then if m is sufficiently large, $x + y$ will coincide with $x + y \bmod m$. In this way, ordinary addition can be accomplished by parallel additions modulo m_i, $i = 1, \ldots, k$. This is the basis of the theory of "fast adders" that are used in computers.

Problems For Section 3.5

1. Find all integer solutions of $18x + 12y = 30$.

2. Find all integer solutions of $11x + 6y = 1$.

3. Find all solutions of $6x \equiv 3 \bmod 9$.

4. Solve the system of congruences $x \equiv 2 \bmod 4$, $x \equiv 1 \bmod 5$, $x \equiv 6 \bmod 9$.

5. Consider Equation (13) in the proof of the Chinese Remainder Theorem. Show that the mapping $(b_1, \ldots, b_k) \to x_0$ is a bijection of $\mathbb{Z}_{m_1} \times \cdots \times \mathbb{Z}_{m_k}$ and $\mathbb{Z}_m$.

3.6 Theorems of Euler and Fermat

In (2.3.3) we met the *Euler phi function*

$\varphi(n) = $ the number of integers in $\{1, \ldots, n\}$ that are relatively prime to n.

If the prime factorization of n is $p_1^{e_1} p_2^{e_2} \ldots p_k^{e_k}$, then by (2.3.4),

$$\varphi(n) = n \left(1 - \frac{1}{p_1}\right) \left(1 - \frac{1}{p_2}\right) \cdots \left(1 - \frac{1}{p_k}\right).$$

In particular, if p is prime, then $\varphi(p^e) = p^e(1 - \frac{1}{p}) = p^e - p^{e-1}$, so $\varphi(p) = p - 1$. The Euler phi function gives us very useful information about inverses $\mod n$.

3.6.1 Euler's Theorem. *If $n \geq 2$ and a is relatively prime to n, then $a^{\varphi(n)} \equiv 1 \mod n$.*

Proof. Let $x_1, x_2, \ldots, x_k$, $k = \varphi(n)$, be the integers between 1 and n that are relatively prime to n. By (3.4.5 (b)), each ax_i is relatively prime to n. Furthermore, if $ax_i \equiv ax_j \mod n$, then by (3.4.5 (c)), $x_i \equiv x_j \mod n$. Thus, modulo n, $ax_1, \ldots, ax_n$ is a permutation of $x_1, \ldots, x_k$. Therefore,

$$(ax_1)(ax_2)\cdots(ax_k) \equiv x_1 x_2 \cdots x_k \mod n,$$

that is,

$$x_1 x_2 \cdots x_k a^{\varphi(n)} \equiv x_1 x_2 \cdots x_k \mod n.$$

By (3.4.5 (b)), $x_1 x_2 \cdots x_k$ is relatively prime to n, so by (3.4.5 (c)), $a^{\varphi(n)} \equiv 1 \mod n$. ∎

3.6.2 Corollary (Fermat's Theorem). *If p is prime and a is any integer, then*
(a) $a^p \equiv a \mod p$
(b) *If p does not divide a, then $a^{p-1} \equiv 1 \mod p$.*

Proof. If p divides a, then $a^p \equiv a \equiv 0 \mod p$, so we may assume that p is not a divisor of a, and therefore that a is relatively prime to p. Now (a) follows from (b) upon multiplication of the congruence by a, so it is sufficient to prove (b). But (b) follows from (3.6.1), since $\varphi(p) = p - 1$. ∎

Theorem 3.6.2 is sometimes called **Fermat's Little Theorem** to emphasize that it is not **Fermat's Last Theorem**.

It is possible to give a direct proof of Fermat's Theorem, without relying on Euler's Theorem. The approach is based on the *binomial expansion modulo p*. If x and y are arbitrary integers and p is prime, then

$$(x + y)^p \equiv x^p + y^p \mod p. \tag{1}$$

To prove this, consider the standard binomial expansion:

$$(x + y)^p = x^p + px^{p-1}y + \binom{p}{2}x^{p-2}y^2 + \cdots + \binom{p}{p-2}x^2 y^{p-2} + pxy^{p-1} + y^p. \tag{2}$$

Now,

$$\binom{p}{r} = \frac{p!}{r!(p-r)!} = \frac{p(p-1)\cdots(p-r+1)}{r!}, \quad r = 1, 2, \ldots, p-1,$$

hence

$$p(p-1)\cdots(p-r+1) = r!\binom{p}{r} = (1)(2)\cdots(r)\binom{p}{r}, \quad r = 1, 2, \ldots, p-1.$$

Therefore p divides $(1)(2) \cdots (r)\binom{p}{r}$, and since p cannot divide any of the factors 1, $2, \ldots, r$ (since $r < p$), p must divide $\binom{p}{r}$ by (3.2.3). Thus every term in the binomial expansion (2) is congruent to $0 \bmod p$, except for x^p and y^p; this proves (1).

Fermat's Theorem may now be proved by induction. We certainly have $0^p \equiv 0 \bmod p$, and if $k^p \equiv k \bmod p$, then $(k+1)^p \equiv k^p + 1^p \equiv k + 1 \bmod p$.

We're not quite finished; what about negative integers? Fortunately, the same approach works: if $k^p \equiv k \bmod p$, then $(k-1)^p \equiv k^p + (-1)^p \equiv k - 1 \bmod p$.

We're still not quite finished; $(-1)^p$ is $+1$ rather than -1 if p is even. But the only even prime is $p = 2$, and modulo 2, -1 is the same as $+1$.

We have now proved part (a) of Fermat's Theorem, and (b) follows from (a) upon dividing the congruence by a. This is legal because if p does not divide a, then a is relatively prime to p, and (3.4.5 (c)) applies.

Problems For Section 3.6

1. Compute (a) $\varphi(600)$, (b) $\varphi(841)$, and (c) $\varphi(6174)$.

2. The integers between 1 and 9 that are relatively prime to 9 are 1, 2, 4, 5, 7, 8. Let $a = 14$, which is relatively prime to 9. Multiply 14 by the numbers 1, 2, 4, 5, 7, 8 in turn, and compute residues $\bmod 9$. Verify that the result is a permutation of 1, 2, 4, 5, 7, 8.

3. Show that the Euler phi function is *multiplicative,* that is, $\varphi(1) = 1$ and if $m, n > 1$ with $\gcd(m, n) = 1$, then $\varphi(mn) = \varphi(m)\varphi(n)$.

4. Give an example to show that if n is composite and $1 < r < n$, n need not divide $\binom{n}{r}$.

5. Show that Fermat's Theorem can be used to compute inverses in $\mathbb{Z}_p$. Is the technique practical?

6. Let $p_1, \ldots, p_k$ be distinct primes, each greater than 2. If $n = 2^{(p_1-1)\cdots(p_k-1)} - 1$, use Fermat's Theorem to show that the product $p_1 \cdots p_k$ divides n.

3.7 The Möbius Inversion Formula

In this section, we derive the formula for computing the Euler phi function by an indirect approach. The basic idea is that if f and g are integer-valued functions defined on the positive integers, and

$$g(n) = \sum_{d \mid n} f(d), \quad n = 1, 2, \ldots \tag{1}$$

where the sum is over all positive divisors of n, then (1) may be solved for f in the sense that f can be expressed in terms of g in a reasonably explicit way. First, we will show how to solve (1). Next, we show that the Euler phi function satisfies (1) with $f(n) = \varphi(n)$ and $g(n) = n$. Finally, we apply the results to obtain the explicit formula for $\varphi(n)$. Our objective is not simply to produce a more complicated derivation of the formula for $\varphi(n)$, but to illustrate a typical application of the Möbius Inversion Formula, which appears in many areas of mathematics.

As usual, $\mathbb{Z}$ will denote the set of all integers, and in this section, the set $\{1, 2, \ldots\}$ of positive integers will be denoted by $\mathbb{Z}^+$.

3.7.1 Definition. The *Möbius function* $\mu : \mathbb{Z}^+ \to \mathbb{Z}$ is defined as follows.

$\mu(1) = 1$, and if $n > 1$, let the prime factorization of n be $p_1^{e_1} \ldots p_k^{e_k}$. If $e_i > 1$ for some i, then $\mu(n) = 0$, and if $e_i = 1$ for all i, then $\mu(n) = (-1)^k$. Thus,

$$\mu(n) = \begin{cases} 1 & \text{if } n \text{ is a product of an even number of distinct primes,} \\ -1 & \text{if } n \text{ is a product of an odd number of distinct primes,} \\ 0 & \text{if } n \text{ contains a repeated prime factor.} \end{cases}$$

For example, $\mu(7) = -1$, $\mu(15) = \mu(3(5)) = 1$, $\mu(45) = \mu(3^2(5)) = 0$. Notice that if $n > 1$, then $\mu(n)$ is nonzero if and only if n is *square-free*, i.e., not divisible by a perfect square greater than 1.

3.7.2 Theorem.

$$\sum_{d|n} \mu(d) = \begin{cases} 0 & \text{if } n \neq 1, \\ 1 & \text{if } n = 1 \end{cases}$$

Proof. First consider the typical example $n = 2^2(3)(5^3)$. If $d|n$, then $\mu(d) \neq 0$ if and only if $d = 1$ or d is a product of one, two or three of the distinct primes 2, 3, 5. Thus the divisors d of n such that $\mu(d) \neq 0$ are $d = 1, 2, 3, 5, 2(3), 2(5), 3(5), 2(3)(5)$, and

$$\sum_{d|n} \mu(d) = 1 - 3 + 3 - 1 = 0.$$

In general, if the distinct prime factors of n are $p_1, \ldots, p_k$, there are $\binom{n}{r}$ divisors that can be formed using products of exactly r of the p_i, and each of these products yields $\mu(d) = (-1)^r$. It is convenient to regard $d = 1$ as a product of none of the p_i; then the contribution of $\mu(1)$ is taken care of by $r = 0$. Therefore,

$$\sum_{d|n} \mu(d) = \sum_{r=0}^{n} \binom{n}{r}(-1)^r,$$

which is the binomial expansion of $(1 - 1)^n = 0$. $\blacksquare$

We are now able to solve Equation (1) for f. The proof to follow contains a tricky reversal of the order of a summation. Suppose we have

$$\sum_{i=1}^{n} \sum_{j=1}^{i} a_{ij};$$

if we blindly reverse the order to obtain

$$\sum_{j=1}^{i} \sum_{i=1}^{n} a_{ij},$$

we get nonsense. The outer index j should range over a fixed set, not one that depends on the inner index i. We can rescue the computation by recognizing that in the original summation we have $1 \le j \le i$, so the correct reversal is

$$\sum_{j=1}^{n} \sum_{i=j}^{n} a_{ij}.$$

This idea should be familiar to you from integral calculus. Think about integrating a function $f(x, y)$ over the triangular region $0 \le x \le 1$, $0 \le y \le x$, and set it up in two ways, integrating first with respect to y and then x, and second with respect to x and then y.

3.7.3 Möbius Inversion Formula. If $f : \mathbb{Z}^+ \rightarrow \mathbb{Z}$, $g : \mathbb{Z}^+ \rightarrow \mathbb{Z}$, and $g(n) = \sum_{d|n} f(d)$, $n = 1, 2, \ldots$, then

$$f(n) = \sum_{d|n} \mu(n/d) g(d)$$

or equivalently,

$$f(n) = \sum_{d|n} \mu(d) g(n/d) = \sum \{ \mu(d') g(d) : dd' = n \}.$$

Before giving the proof, let's look at a concrete example to see how the Möbius function actually accomplishes the inversion. If $n = 12$, we hope to show that

$$f(12) = \mu(12)g(1) + \mu(6)g(2) + \mu(4)g(3) + \mu(3)g(4) + \mu(2)g(6) + \mu(1)g(12),$$

that is,

$$f(12) = 0g(1) + 1g(2) + 0g(3) - 1g(4) - 1g(6) + 1g(12)$$
$$= g(2) - g(4) - g(6) + g(12).$$

To check that this is correct, note that by definition of g,

$$g(2) = f(1) + f(2)$$
$$g(4) = f(1) + f(2) \qquad\quad + f(4)$$
$$g(6) = f(1) + f(2) + f(3) \qquad\quad + f(6)$$
$$g(12) = f(1) + f(2) + f(3) + f(4) + f(6) + f(12).$$

When we compute $g(2) - g(4) - g(6) + g(12)$, the terms $f(1)$, $f(2)$, $f(3)$, $f(4)$, and $f(6)$ cancel, leaving only $f(12)$, as desired.

Proof of the Möbius Inversion Formula. By hypothesis,

$$\sum_{d|n} \mu(n/d) g(d) = \sum_{d|n} \mu(n/d) \sum_{e|d} f(e),$$

and we know that we must be careful when reversing the order of summation. The index e depends on the index d, but if e divides d and d divides n, then e divides n. Thus we may reverse the order as follows:

$$\sum_{d|n} \mu(n/d)g(d) = \sum_{e|n} f(e) \sum \{\mu(n/d) : e|d \text{ and } d|n\} \tag{2}$$

The inner summation on the right looks formidable, but let's see if we can decode it. If $e|d$ and $d|n$, then n/d divides n/e; to see this, write $n = ad$, $d = be$, and note that $\frac{n}{e} = \frac{n}{d}\frac{d}{e} = \frac{n}{d}b$. So we are summing $\mu(d')$ where d' ranges over some of the divisors of n/e. In fact all the divisors of n/e are represented. For suppose that e divides n and d' divides n/e. Then d' divides n, so $n = dd'$ for some divisor d of n. We claim that $e|d$; to verify this, note that $\frac{n}{e} = qd'$ for some positive integer q, hence $d = eq$. Finally, $n/d = n/eq = d'$. Consequently,

$$\sum \{\mu(n/d) : e|d \text{ and } d|n\} = \sum \left\{ \mu(d') : d' \text{ divides } \frac{n}{e} \right\}$$
$$= \begin{cases} 1 & \text{if } e = n, \\ 0 & \text{if } e \neq n \end{cases} \tag{3}$$

by (3.7.2). Therefore by (2) and (3),

$$\sum_{d|n} \mu(n/d)g(d) = f(n). \blacksquare$$

Remark. It is tempting to shorten the above proof by skipping directly from (2) to (3). But then the question "How do you know that d' ranges over all divisors of n/e?" is a natural one, and to answer the question, some hard, slogging work is needed.

An additional property of the Möbius function will be useful.

3.7.4 Theorem. *If the prime factorization of the integer $n \geq 2$ is $n = p_1^{e_1} p_2^{e_2} \ldots p_k^{e_k}$, then*

$$\left(1 - \frac{1}{p_1}\right)\left(1 - \frac{1}{p_2}\right) \cdots \left(1 - \frac{1}{p_k}\right) = \sum_{d|n} \frac{\mu(d)}{d}. \tag{4}$$

Proof. Expand the left side of (4) to obtain

$$1 - \frac{1}{p_1} - \frac{1}{p_2} - \cdots - \frac{1}{p_k} + \frac{1}{p_1 p_2} + \frac{1}{p_1 p_3}$$
$$+ \cdots + \frac{1}{p_{k-1} p_k} - \cdots + (-1)^k \frac{1}{p_1 p_2 \ldots p_k}. \tag{5}$$

As in the proof of (3.7.2), the denominators of the terms of (5) are precisely the divisors d of n such that $\mu(d) \neq 0$. Furthermore, if an even number of primes appear in the denominator d, then $\mu(d) = +1$, and if an odd number of primes appear, then $\mu(d) = -1$. Thus (5) coincides with

$$\sum_{d|n} \frac{\mu(d)}{d}. \blacksquare$$

The next result allows the Möbius Inversion Formula to be applied to the Euler phi function.

3.7.5 Theorem.   *For every positive integer n we have $\sum_{d|n} \varphi(d) = n$.*

Proof.   First let's check the result for the specific case $n = 15$. The divisors of 15 are 1, 3, 5, and 15, and

$$\varphi(1) + \varphi(3) + \varphi(5) + \varphi(15) = 1 + 2 + 4 + 8 = 15,$$

as asserted. To see why this happens, write out all the fractions x/n, $x = 1, 2, \ldots, n$:

$$\frac{1}{15} \; \frac{2}{15} \; \frac{3}{15} \; \frac{4}{15} \; \frac{5}{15} \; \frac{6}{15} \; \frac{7}{15} \; \frac{8}{15} \; \frac{9}{15} \; \frac{10}{15} \; \frac{11}{15} \; \frac{12}{15} \; \frac{13}{15} \; \frac{14}{15} \; \frac{15}{15}.$$

Reduce all fractions to lowest terms to obtain

$$\frac{1}{15} \; \frac{2}{15} \; \frac{1}{5} \; \frac{4}{15} \; \frac{1}{3} \; \frac{2}{5} \; \frac{7}{15} \; \frac{8}{15} \; \frac{3}{5} \; \frac{2}{3} \; \frac{11}{15} \; \frac{4}{5} \; \frac{13}{15} \; \frac{14}{15} \; \frac{1}{1}.$$

The denominators on this list are the divisors d of n ($d = 1, 3, 5, 15$ in this case). For each d, the fraction e/d will appear on the list if and only if e is relatively prime to d. For example, $4/15$ appears, but $6/15$ does not; $6/15$ reduces to $2/5$, which is associated with $d = 5$ rather than $d = 15$. Thus, for each d there are $\varphi(d)$ terms on the list, and since the total number of terms is n, we have $\sum_{d|n} \varphi(d) = n$. ∎

We can now find the explicit formula for $\varphi(n)$. Applying the Möbius Inversion Formula (3.7.3) to (3.7.5), we get

$$\varphi(n) = \sum_{d|n} \mu(d) \frac{n}{d} = n \sum_{d|n} \frac{\mu(d)}{d}.$$

If $n = p_1^{e_1} \cdots p_k^{e_k}$, then by (3.7.4),

$$\varphi(n) = n \left(1 - \frac{1}{p_1} \right) \cdots \left(1 - \frac{1}{p_k} \right)$$

as found earlier.

Problems For Section 3.7

1.   Show (with only modest effort) that $\mu(17305893) = 0$.

2.   Show that the Möbius function is multiplicative, that is, $\mu(1) = 1$ and if $m, n > 1$ with $\gcd(m, n) = 1$, then $\mu(mn) = \mu(m)\mu(n)$.

3.   Let $f : \mathbb{Z}^+ \to \mathbb{Z}$ be a multiplicative function, and define $g(n) = \sum_{d|n} f(d)$ as in the Möbius Inversion Formula. Show that g is also multiplicative. [If $n = p_1^{e_1} \cdots p_k^{e_k}$, show that

$$g(n) = (1 + f(p_1) + f(p_1^2) + \cdots + f(p_1^{e_1})) \cdots (1 + f(p_k) + f(p_k^2) + \cdots + f(p_k^{e_k}))].$$

4.   Let $S_r(n) = \sum_{d|n} d^r$, $r, n = 1, 2, \ldots$. Show that each S_r is multiplicative.

5.   Define a positive integer n to be *perfect* if n is the sum of its positive divisors, with the exception of n itself. Show that n is perfect if and only if $S_1(n) = 2n$ (where S_1 is defined in Problem 4).

6. Let n be a positive integer. If $2^n - 1$ is prime, show that $2^{n-1}(2^n - 1)$ is perfect (and even). [If $x = 2^{n-1}(2^n - 1)$, then $S_1(x) = S_1(2^{n-1})S_1(2^n - 1)$ by Problem 4. Compute $S_1(2^{n-1})$ and $S_1(2^n - 1)$ directly, and verify that $S_1(x) = 2x$.]

7. Let x be a perfect even number, so that $x = 2^h q$ with $h \geq 1$ and q odd.

(a) Show that $2x = (2^{h+1} - 1)S_1(q)$. [Use Problems 4 and 5.]

(b) Show that $S_1(q) > q$. [Since $2x = 2(2^h q) = 2^{h+1}q$, part (a) yields $2^{h+1}q = (2^{h+1} - 1)S_1(q)$.]

(c) Let $S_1(q) = q + r$, where $r > 0$ by part (b). Show that $q = (2^{h+1} - 1)r$.

(d) Show that in part (c), $r = 1$, hence $q = 2^{h+1} - 1$. [Since $h \geq 1$, we have $2^{h+1} - 1 > 1$, so by (c), r is a divisor of q with $r < q$. If $r > 1$, show that $S_1(q) > q + r$, which contradicts (c).]

(e) Show that q is prime. [Use (c) and (d).]

Conclude that x is a perfect even number if and only if x is of the form $2^{n-1}(2^n - 1)$, where $2^n - 1$ is prime.

4

Some Highly Informal Set Theory

Set theory is a fundamental area of mathematics and makes contact with all other areas. A mathematician must know how to use Zorn's Lemma, and be familiar with the arithmetic of cardinal numbers. A formal development of the required set theory takes considerable time, so I will try to give an intuitive development that suggests why results are valid. This does not replace a course in set theory, and I would urge you to take such a course at some point, if you have not done so already.

4.1 Well-Orderings

In Section 1.6 we introduced partial orderings, and we now study these relations in more detail. We will use $\leq$ rather than the generic symbol R to denote a partial ordering, even though the given relation need not have any connection with "less than or equal to" on numbers. Recall that a partial ordering on a set A is a relation on A that is

reflexive: $a \leq a$ for all $a \in A$;
antisymmetric: if $a \leq b$ and $b \leq a$ then $a = b$; and
transitive: if $a \leq b$ and $b \leq c$ then $a \leq c$.

A *total ordering* is a partial ordering in which any two elements can be compared: either $a \leq b$ or $b \leq a$. We will be interested in a special class of total orderings.

4.1.1 Definition. A *well-ordering* on a set A is a total ordering in which every nonempty subset S of A has a smallest element relative to the ordering. In other words, there is an element $s \in S$ such that $s \leq x$ for all $x \in S$.

If $\mathbb{N}$ is the set $\{0, 1, 2, \ldots\}$ of nonnegative integers (also called *natural numbers*), then less than or equal to on $\mathbb{N}$ is a well-ordering. One method of proof uses *strong induction,* which is a variation of ordinary mathematical induction. Suppose that we are trying to prove that a statement P_n holds for all $n = 0, 1, \ldots$. (We start at $n = 0$ because we will be working with natural numbers.) A proof by strong induction goes like this:

1. Prove the *basis* step P_0.

2. Assuming that $P_0, P_1, \ldots, P_n$ are *all* valid (the *strong induction hypothesis*), prove that P_{n+1} is true.

It follows that P_n holds for all n. We know that P_0 is true, and if P_0 holds, so does P_1. But now we have both P_0 and P_1, and we can conclude P_2. Similarly, from P_0, P_1, and P_2 we can establish P_3, and so on.

Here is the proof by strong induction.

4.1.2 Theorem. *Less than or equal to on the set $\mathbb{N}$ of natural numbers is a well-ordering.*

Proof. Let S be a subset of $\mathbb{N}$. If S is nonempty, we must show that S has a smallest element. We'll prove the contrapositive, namely that if S has no smallest element, then S is empty. We will show that for every $n = 0, 1, \ldots$, n does not belong to S. Thus our statement P_n is "$n \notin S$". Now $0 \notin S$, for if 0 were in S, then it would certainly be the smallest member of S. Assuming that none of the integers $0, 1, \ldots, n$ belongs to S, we claim that $n+1 \notin S$. Indeed, if $n+1$ were in S, then since $0, 1, \ldots, n \notin S$, $n+1$ would be the smallest element of S. By strong induction, S is empty. ∎

There is an alternative argument which is perhaps more intuitive. If S is a nonempty subset of $\mathbb{N}$, we must produce a smallest element of S. Since S is nonempty, there is some number $n_1 \in S$. If n_1 is the smallest member of S, we are finished; if not, there is an element $n_2 \in S$ with $n_2 < n_1$. If n_2 is the smallest member of S, we are finished; if not, there is an element $n_3 \in S$ with $n_3 < n_2$, and so on. Since there are only finitely many natural numbers less than n_1, this process must produce a smallest element in a finite number of steps.

If we consider less than or equal to on the entire set $\mathbb{Z}$ of integers, or on the rational numbers $\mathbb{Q}$, or on the real numbers $\mathbb{R}$, we do *not* have a well-ordering. For there is no smallest integer, no smallest rational number, and no smallest real number. However, $\mathbb{Z}$ *can* be well-ordered in the following way. Write the elements of $\mathbb{Z}$ in the following order: $0, 1, -1, 2, -2, 3, -3, \ldots$. This places $\mathbb{Z}$ in one-to-one correspondence with $\mathbb{N}$, and uses the standard well-ordering of $\mathbb{N}$ to impose a well-ordering on $\mathbb{Z}$. A similar procedure works on the rationals (see Section 2.4), but fails when we try to apply it to the uncountable set $\mathbb{R}$. Can we well-order the reals? Nobody has come up with an explicit description, but there is the following axiom of set theory, known as the **Well-Ordering Principle.**

4.1.3 Axiom. *Every set can be well-ordered.*

Why should we build such an extravagant axiom into our theory? Here is one possible argument. Suppose that S is an arbitrary set; how can we well-order S? If S is nonempty, pick any element of S and label it a_0; then if possible, pick a different element and label it a_1, and continue in this fashion. If the process does not terminate, then we have $a_0, a_1, a_2, \ldots$ in S, so that S essentially contains a copy of the natural numbers. Now if possible, pick a new element of S and label it b_0, then another new element b_1, and continue as before. Keep going in this fashion until S is exhausted.

The problem is that the last sentence does not in any way describe an algorithm that can be implemented by a computer program. Yet it seems reasonable and has led to some very interesting mathematics. So the Well-Ordering Principle has been accepted by most (but not all) mathematicians.

Let's come back to an example we considered in Section 1.6, the divisibility relation on the set $A = \{1, 2, \ldots, 12\}$ consisting of the first 12 positive integers. The subset $B = \{1, 3, 6\}$ of A is a *chain* (totally ordered subset) of A, for if $a, b \in B$, then either a divides b or b divides a. Now a *maximal chain* is one that is not contained in a strictly larger chain. Is B a maximal chain? No, because $B \subset C = \{1, 3, 6, 12\}$. But if we try to enlarge C, we destroy the chain property. Thus $\{1, 3, 6, 8, 12\}$ is not a chain, because, for example, 3 and 8 cannot be compared; 3 does not divide 8 and 8 does not divide 3. Therefore C is a maximal chain.

In a finite partially ordered set, any chain is contained in a maximal chain. We can start with the given chain and try to enlarge it by adding elements one at a time until we can go no farther. This natural procedure essentially carries over to an arbitrary partially ordered set, if we have the well-ordering principle available.

Let A be a partially ordered set, and let B be any chain of A. First, we well-order the set $A \setminus B$; for convenience, write $A \setminus B = \{a_i : i \in I\}$ where I is a well-ordered set. Since I keeps track of the indices i of the elements a_i, it is called an *index set*. We are going to start with B and create larger and larger sets until we reach our goal, a maximal chain. The technique has a fancy name—

4.1.4 Definition by Transfinite Induction.   The idea is to carry out inductive procedures (see (1.6.5)) on arbitrary well-ordered sets, not simply on the natural numbers. Here is how the process works to yield a maximal chain. [Notation: $\cup\{B_j : j < i\}$ means the union of all the sets B_j for indices j less than i.]

We set $B_0 = B$; having defined B_j for all $j < i$, we define B_i as follows. If a_i is comparable to all elements in all the B_j, $j < i$, then $\cup\{B_j : j < i\} \cup \{a_i\}$ forms a chain, and we set $B_i = \cup\{B_j : j < i\} \cup \{a_i\}$. Otherwise we take $B_i = \cup\{B_j : j < i\}$. All we are saying is that if a_i can be added to all the previous sets without destroying the chain property, we add it; otherwise, we don't. The validity of the definition of the B_i follows from the Well-Ordering Principle. For if B_i fails to be well-defined for some i, there is a smallest such i. But then B_j is well-defined for all $j < i$, and the explicit definition above shows that B_i is then well-defined, which is a contradiction.

We may now produce a maximal chain containing B.

4.1.5 Maximum Principle.   *Let A be a partially ordered set, and let B be any chain of A. Then B is contained in at least one maximal chain of A.*

Proof.   Define the sets B_i as above, and set $C = \cup\{B_i : i \in I\}$. We claim that C is a maximal chain. For an element not belonging to C can't belong to the subset B, so it must be one of the a_i. Now when we reach i in the inductive process, we must choose not to enlarge the chain. (If we do enlarge the chain, then $a_i \in B_i \subseteq C$.) But then a_i is not comparable to at least one element of some B_j, $j < i$, and hence is not comparable to at least one element of C. In other words, there is no way to adjoin a_i to C and have the resulting set remain a chain. Thus no set strictly containing C can be a chain, and consequently C is maximal. ∎

In proving the Maximum Principle, we made use of a definition by transfinite induction. It is also possible to do proofs by transfinite induction, as follows.

4.1.6 Proof by Transfinite Induction.   Suppose we wish to prove that the statement P_i holds for all i in the well-ordered set I. A proof by transfinite induction is analogous to an argument via strong induction:

1.   Prove the *basis step* P_0, where 0 is the smallest element of I.

2.   Let $i > 0$. Assuming that P_j is valid for all $j < i$ (the *transfinite induction hypothesis*), prove that P_i holds.

It then follows that P_i is true for all $i \in I$. For if not, there is a smallest index i for which P_i is false, and $i > 0$ by the basis step. But then P_j is true for all $j < i$, and by the transfinite induction hypothesis, P_i is true, a contradiction.

Problems For Section 4.1

1.   Again consider the divisibility relation on $A = \{1, 2, \ldots, 12\}$. If $B = \{1, 2\}$, find two distinct maximal chains containing B.

2.   Let A and B be well-ordered sets, and partially order $A \times B$ as follows:

$$(a_1, b_1) \le (a_2, b_2) \quad \text{if and only if} \quad a_1 < a_2 \text{ (that is, } a_1 \le a_2 \text{ but } a_1 \ne a_2)$$

$$\text{or} \quad a_1 = a_2 \text{ and } b_1 \le b_2.$$

This is called *lexicographic (dictionary) ordering*. Show that lexicographic ordering well-orders $A \times B$.

3.   In Problem 2, suppose we change the definition of the ordering to

$$(a_1, b_1) \le (a_2, b_2) \quad \text{if and only if} \quad a_1 \le a_2 \text{ and } b_1 \le b_2.$$

Do we still have a well-ordering of $A \times B$?

4.2 Zorn's Lemma and the Axiom of Choice

The Maximum Principle is a result about maximal chains in a partially ordered set. We now shift the focus to maximal *elements* rather than chains. Recall that to say that the element a in the partially ordered set A is maximal means that there is no element $b \in A$ such that $a < b$. Every finite partially ordered set has at least one maximal element, as we found in (1.6.5). But there are many infinite partially ordered sets without a maximal element; for example, consider less than or equal to on the integers. Zorn's Lemma establishes the existence of a maximal element under certain conditions. First, some terminology—

4.2.1 Definition.   Let A be partially ordered by $\leq$, and let B be a subset of A. The element x is an *upper bound* of B in A if $x \in A$ and $b \leq x$ for all $b \in B$; note that x need not belong to B.

Let's look again at the integers under less than or equal to. If n is any integer, then the set $\{n, n+1, n+2, \ldots\}$ is a chain with no upper bound. If we eliminate this possibility by requiring every chain to have an upper bound, then maximal elements always exist.

4.2.2 Zorn's Lemma.   *Let A be a nonempty partially ordered set such that every chain of A has an upper bound in A. Then A has a maximal element.*

Proof.   Since A is nonempty, it has an element a. Apply the Maximum Principle (4.1.5) to the chain $\{a\}$ to conclude that A has a maximal chain C. By hypothesis, C has an upper bound $x \in A$. If $x \notin C$, we can make a larger chain $C \cup \{x\}$; we still have a chain because $c \leq x$ for every $c \in C$. This contradicts the maximality of C, and consequently $x \in C$. In fact x is a maximal element of A, for if $x < y \in A$, then for every $c \in C$ we have $c \leq x < y$, so that $C \cup \{y\}$ is a chain of A, contradicting the maximality of C. ∎

The proof of Zorn's Lemma depends on the Maximum Principle, which in turn is a consequence of the Well-Ordering Principle. Here is another application of the Well-Ordering Principle.

4.2.3 Axiom of Choice.   *If we have a family of nonempty sets A_i, $i \in I$, we can choose an element in each A_i. More precisely, there is a function f whose domain is I such that for each $i \in I$, we have $f(i) \in A_i$.*

Proof.   Well-order each of the sets A_i, and let a_i be the smallest element of A_i. Then take $f(i) = a_i$. ∎

The Well-Ordering Principle, the Maximum Principle, Zorn's Lemma, and the Axiom of Choice are *equivalent* in the sense that if any one of these statements is added to the basic axioms of set theory, then all the other statements can be deduced. The proof of 4.2.2 shows that the Maximum Principle implies Zorn's Lemma. Problem 1 asks you to show that conversely, Zorn's Lemma implies the Maximum Principle.

Problems 2–5 give a preview of the use of Zorn's Lemma in algebra. Euclidean three-dimensional space is an example of a *vector space* V over a field F (F is the field of real numbers in this case.) The elements (vectors) in V can be added and subtracted, that is, V is an abelian group under addition. Vectors can be multiplied by scalars; in other words, for each $v \in V$ and $r \in F$ there is an element $rv \in V$, and the following properties are satisfied:

$$r(x + y) = rx + ry, \quad (r + s)x = rx + sx, \quad r(sx) = (rs)x, \quad 1x = x$$

for all $x, y \in V$ and $r, s \in F$.

Let V be an arbitrary vector space. The elements $v_1, \ldots, v_n \in V$ are said to be *linearly independent* if none of the vectors can be expressed as a linear combination of the others. We also say that the set $\{v_1, v_2, \ldots, v_n\}$ is linearly independent. If the v_i are not linearly independent, so that one of the vectors can be expressed as a linear combination of the others, the v_i are said to be *linearly dependent* (and the set $\{v_1, \ldots, v_n\}$ is linearly dependent). An arbitrary (possibly infinite) set S of vectors is said to be *linearly independent* if none of the vectors in S can be expressed as a finite linear combination of other vectors in S; otherwise, S is *linearly dependent.*

(A formal detail must be attended to if $n = 1$. We regard $\{v_1\}$ as linearly independent if and only if v_1 is not the zero vector.)

Problems 2–5 yield the result that every vector space V has a maximal linearly independent set; such a set is called a *basis* for V.

Problems For Section 4.2

1. Show that the Maximum Principle is a consequence of Zorn's Lemma.

2. Let $v_1, \ldots, v_n$ be elements of the vector space V. Show that $v_1, \ldots, v_n$ are linearly independent if and only if

$$r_1 v_1 + \cdots + r_n v_n = 0 \quad \text{implies} \quad r_1 = \cdots = r_n = 0.$$

3. Let S be an arbitrary subset of the vector space V. Show that S is linearly independent if and only if for all $n = 1, 2, \ldots$, all $v_1, \ldots, v_n \in S$ and all $r_1, \ldots, r_n \in F$,

$$r_1 v_1 + \cdots + r_n v_n = 0 \quad \text{implies} \quad r_1 = \cdots = r_n = 0.$$

4. Consider the collection $\mathcal{C}$ of all linearly independent subsets of V, partially ordered by inclusion; that is, $A \leq B$ means that $A \subseteq B$. Note that $\mathcal{C}$ is nonempty since any set $\{v\}$ where $v \neq 0$ is linearly independent. Show that every chain of $\mathcal{C}$ has an upper bound in $\mathcal{C}$.

5. Use Zorn's Lemma to show that every vector space V has a *basis,* that is, a maximal linearly independent set.

4.3 Cardinal Numbers

We are going to try to determine when one infinite set is larger than another. Suppose we have two sets A and B, and an injective mapping $f : A \to B$. Then f identifies A with the subset $f(A)$ of B, and possibly there are points of B left over. So it is reasonable to conclude that A is no larger than B. If there is a bijection, in other words, a one-to-one correspondence, between A and B, it is reasonable to say that A and B have the same size. Let's start the mathematical machinery.

4.3.1 Definitions and Comments. Define $A \leq_s B$ if there is an injective $f : A \to B$; we say that the size of A is less than or equal to that of B. If there is a bijection between A and B, we say that A and B have the same size, and write $A =_s B$.

As it stands, $\leq_s$ is not a partial ordering. For example, $\{a, b\} \leq_s \{c, d\}$ and $\{c, d\} \leq_s \{a, b\}$, but $\{a, b\} \neq \{c, d\}$. But from the point of view of counting elements, there really is no difference between $\{a, b\}$ and $\{c, d\}$, so let's regard the two sets as the same. In general, we identify sets of the same size. Mathematically, we are forming an equivalence relation: $A \sim B$ if A and B have the same size, and we are working with equivalence classes of sets rather than the individual sets themselves. But this viewpoint probably interferes with the learning process at this stage, so let's consider the relation $\leq_s$ on sets, regarding two sets of the same size as essentially the same set.

Let's try to prove that $\leq_s$ is a partial ordering. Certainly it is reflexive; if A is any set, let $f : A \to A$ be the *identity function*; that is, $f(x) = x$ for all $x \in A$. Since f is injective, we have $A \leq_s A$. Transitivity is also not a problem; if $f : A \to B$ and $g : B \to C$ are injective, let h be the composition $g \circ f$. Then $h : A \to C$ and h is injective, for if $g\big(f(x)\big) = g\big(f(y)\big)$, then $f(x) = f(y)$ since g is injective, and therefore $x = y$ since f is injective. Thus if $A \leq_s B$ and $B \leq_s C$, then $A \leq_s C$.

To investigate antisymmetry, we need a major theorem, which we prove after introducing some terminology.

4.3.2 Definitions. If $f : A \to B$ and C is a subset of A, the *restriction* of f to C is the mapping $f_C : C \to B$ defined by $f_C(x) = f(x)$, $x \in C$. In other words, we cut the domain of f down from A to C.

If $f : A \to B$ is injective, the *inverse* of f is the function $f^{-1} : f(A) \to A$ defined by $f^{-1}\big(f(x)\big) = x$ for every $f(x) \in f(A)$. In other words, f^{-1} maps backwards: if $f(x) = y$, then $f^{-1}(y) = x$. If there is an arrow from x to y in a diagram describing f, then there will be an arrow from y to x in the diagram that pictures f^{-1}. If f is bijective, then the domain of f^{-1} is all of B.

In Section 1.5, we used the same symbol f^{-1} to describe preimages: $f^{-1}(C)$ is the preimage of the subset $C \subseteq B$ under the mapping f. If there is any chance of confusion, we will say explicitly which usage is intended.

4.3.3 Schröder-Bernstein Theorem. *If there is an injective mapping $f : A \to B$ and an injective mapping $g : B \to A$, then there is a bijection $h : A \to B$.*

Proof. Start with $a \in A$ and trace its ancestry. We say that a has a parent b if $b \in B$ and $g(b) = a$. Similarly, b has a parent a' if $a' \in A$ and $f(a') = b$; see Figure 4.3.1.

Note that an element can have at most one parent, because f and g are injective.

In tracing the ancestry of $a \in A$, there are three possibilities.

Case 1. We reach an element in A with no parent, for example, a'' in Figure 4.3.1. In this case, place a in the set A_1.

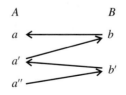

FIGURE 4.3.1
Proof of the Schröder-Bernstein Theorem

Case 2. We reach an element in B with no parent. This would happen if in Figure 4.3.1, the arrow from a'' to b' were missing; then we would reach b'. In this situation, place a in the set A_2.

Case 3. The process does not terminate; here, we place a in A_3.

Similarly, in tracing the ancestry of $b \in B$, we have the same three cases, and place b in the set B_1, B_2, or B_3 depending on which case occurs.

In Figure 4.3.1, both a and $f(a') = b$ have the same parentless ancestor a'', with a' in A_1 and b in B_1. Now consider the restriction of f to A_1. Since every $b \in B_1$ is of the form $f(a')$ for some $a' \in A_1$, and f is injective, we obtain a bijection between A_1 and B_1. Similarly, the restriction of f to A_3 is a bijection of A_3 and B_3, and the restriction of g to B_2 is a bijection of B_2 and A_2. We now define $h = f$ on $A_1 \cup A_3$, and $h = g^{-1}$ (the inverse of g) on A_2. Then h is the desired bijective mapping between A and B. ∎

4.3.4 Corollary. *If $A \leq_s B$ and $B \leq_s A$, then $A =_s B$. Thus if we call sets of the same size equivalent, then $\leq_s$ on equivalence classes is a partial ordering.*

Proof. The first statement is the Schröder-Bernstein Theorem (4.3.3). If $|A|$ is the equivalence class of A, we take $|A| \leq |B|$ to mean that $A \leq_s B$. This is well defined, because if A' has the same size as A, and B' has the same size as B, then by composing the bijection between A' and A with the injection from A to B and following this map by the bijection between B and B', we establish that $A' \leq_s B'$. The discussion preceding (4.3.2) shows that $\leq$ is reflexive and transitive, and (4.3.3) shows that $\leq$ is antisymmetric. ∎

With the aid of Zorn's Lemma, we can show that any two sets can be compared by size, that is, $\leq$ is a total ordering.

4.3.5 Theorem. *If A and B are arbitrary sets, then either $A \leq_s B$ or $B \leq_s A$.*

Proof. This is a canonical example of the use of Zorn's Lemma in set theory. Consider the collection $\mathcal{C}$ of all triples (A_i, B_i, f_i) where A_i is a subset of A, B_i is a subset of B, and f_i is an injective mapping from A_i to B_i. Call $(A_i, B_i, f_i) \leq (A_j, B_j, f_j)$ if and only if $A_i \subseteq A_j$, $B_i \subseteq B_j$, and the restriction of f_j to A_i is f_i. Every chain has an upper bound (A_0, B_0, f) where A_0 is the union of the A_i in the chain and B_0 is the union of the B_i. If $x \in A_i$ we take $f(x) = f_i(x)$, and by our definition of $\leq$ there can be no ambiguity if x also belongs to A_j for some $j \neq i$. Since each f_i is injective, f must be injective as well. The collection $\mathcal{C}$ is nonempty since it contains $(\varnothing, \varnothing, \varnothing)$. [If you prefer not to work with the empty function, defined by $\varnothing(x) =$ anything you like for $x \in \varnothing$, you may assume that A and B are nonempty. (We have no trouble determining the size of $\varnothing$.) If $a \in A$ and $b \in B$, let g map a to b; then $(\{a\}, \{b\}, g)$ belongs to $\mathcal{C}$.]

By Zorn's Lemma, there is a maximal element (A^*, B^*, f^*). If $A^* \subset A$ and $B^* \subset B$, we have elements $a \notin A^*$ and $b \notin B^*$, and we can map a to b to contradict maximality of (A^*, B^*, f^*). Thus either $A^* = A$, in which case $A \leq_s B$, or $B^* = B$, in which case there is an injective map from A^*, a subset of A, onto B. The inverse of this function is an injective map from B to A^*, so that $B \leq_s A$. ∎

4.3.6 Definitions and Comments. The equivalence class $|A|$ of a set A, that is, the collection of sets having the same size as A, is called the *cardinal number* or *cardinality* of A. The cardinal number of a countably infinite set (such as the integers or the rationals) is denoted by $\aleph_0$ (read aleph null). Usually we are less formal, and simply identify $\aleph_0$ with any countably infinite set that we find convenient. Similarly, we identify $|A|$ with any convenient representative of the equivalence class, for example, A. Thus $|A| \leq |B|$ if and only if $A \leq_s B$.

Given any cardinal number, we can always find a larger one with the aid of the following result.

Let A be any set, and let 2^A (called the *power set* of A) be the set consisting of all subsets of A. The reason for the notation 2^A is that a subset B of A can be identified with a function f from A to a two-element set, say $\{0, 1\}$. Then $f(x) = 1$ indicates that $x \in B$, and $f(x) = 0$ indicates that $x \notin B$. A subset B of A is specified by making a two-valued choice for each x in A, so the total number of subsets is $2 \times 2 \times 2 \times \cdots$ ($|A|$ times). Similarly, a function from A to an arbitrary set C is specified by choosing an element of C for each x in A, so the notation C^A is used for the set of all functions from A to C.

If A has n elements, then 2^A has 2^n elements (Section 2.1, Problem 4). Even when A is infinite, the power set always has strictly larger cardinality.

4.3.7 Theorem. *For any set A, the power set 2^A is of larger size than A.*

Proof. Since the map $a \to \{a\}$ is injective, $A \leq_s 2^A$. Thus we must show that A and 2^A cannot have the same size. Suppose then that $f : A \to 2^A$ is a bijection, and let S be the set of all a in A such that $a \notin f(a)$. Since f is surjective, we have $f(x) = S$ for some $x \in A$. Does x belong to $f(x)$? If it does, then $x \in S$ (because $f(x) = S$). It follows by definition of S that $x \notin f(x)$. Similarly, if $x \notin f(x)$ then $x \notin S$, so that $x \in f(x)$. We have a contradiction, so that no such bijection can exist. ∎

Problems For Section 4.3

1. If $f : A \to B$ and $g : B \to C$ are surjective, show that the composition $g \circ f$ is surjective.

2. Give an example of a function $f : A \to B$ and a subset C of A such that f is not injective, but the restriction f_C of f to C is injective.

3. Show that $B \leq_s A$ if and only if there is a surjective mapping $f : A \to B$.

4. Show that the set B is countable if and only if there is a surjective mapping $f : \mathbb{N} \to B$, where $\mathbb{N}$ is the set of natural numbers.

5. If A is a finite set with m elements, and B is a finite set with n elements, show that $|A| = |B|$ if and only if $m = n$, and $|A| \leq |B|$ if and only if $m \leq n$. (I know this is obvious, but work it out anyway to get some practice using the definition of cardinal number.)

4.4 Addition and Multiplication of Cardinals

If $|A|$ and $|B|$ are cardinal numbers, we would like to define $|A| + |B|$ and $|A||B|$ in a sensible way; here is one possible approach.

4.4.1 Definitions and Comments.   To add $|A|$ and $|B|$, it is reasonable to write down the elements of A and then list the elements of B, and count the elements in the resulting set. A technical difficulty occurs if there is an element belonging to both A and B, and we will legislate this problem out of existence by assuming that A and B are disjoint. But what if they aren't? If $c \in A \cap B$, we can replace c by $(1, c)$ in the list of elements in A, and by $(2, c)$ in the list of elements in B, forcing disjointness. The sizes of A and B are unaffected by this maneuver, so we may as well assume at the beginning that A and B are disjoint. We define

$$|A| + |B| = |A \cup B|.$$

If A and B are finite sets with m and n elements respectively, then the Cartesian product $A \times B$ has mn elements. (If we select an ordered pair (a, b), $a \in A$, $b \in B$, we have m choices for a and n choices for b, and the Multiplication Rule (2.1.1) applies.) For arbitrary cardinal numbers $|A|$ and $|B|$, it is natural to define

$$|A||B| = |A \times B|.$$

If A and B are countably infinite, cardinal arithmetic is very pleasant.

4.4.2 Theorem.
  (a) $\aleph_0 + \aleph_0 = \aleph_0$
  (b) $\aleph_0 \aleph_0 = \aleph_0$

Proof.
  (a) Identify the first $\aleph_0$ with the nonnegative integers $\mathbb{N}$ and the second with the negative integers $\mathbb{Z}^-$. Then (a) says that the union of $\mathbb{N}$ and $\mathbb{Z}^-$, which is the set $\mathbb{Z}$ of all integers, is countably infinite. We verified this at the beginning of Section 2.4.
  (b) Identify the first and the second $\aleph_0$ with the positive integers $\mathbb{Z}^+$. Then (b) says that the set of all ordered pairs (x, y) of positive integers is countably infinite, and this follows from Cantor's Diagonal Process (Section 2.4). ∎

General cardinal arithmetic is almost as simple. In part (b) of the following result, 0 denotes the cardinal number of the empty set.

4.4.3 Theorem.
  (a) *Let α and β be cardinals, with $\alpha \le \beta$ and β infinite. Then $\alpha + \beta = \beta$.*
  (b) *If $\alpha \neq 0$, $\alpha \le \beta$ and β is infinite, then $\alpha\beta = \beta$.*

Proof.
  (a) Recall from our discussion of the Well-Ordering Principle (4.1.3) that β can be decomposed as a union of countably infinite sets. If B_1 is a countably infinite set

occurring in the decomposition of β, and B_2 is a disjoint copy of B_1, then $B_1 \cup B_2$ will occur in the decomposition of $\beta + \beta$. But by (4.4.2(a)), $B_1 \cup B_2$ is countably infinite, so $\beta + \beta = \beta$. Therefore

$$\beta \leq \alpha + \beta \leq \beta + \beta = \beta,$$

and the result follows.

(b) First note that $\beta \leq \alpha\beta$; for if $|B| = \beta$ and $|A| = \alpha$, pick any $a \in A$ (using the hypothesis that $A \neq \varnothing$). The map $b \to (a, b)$ of B into $A \times B$ is injective, so $\beta \leq \alpha\beta$. Next, observe that since $\alpha \leq \beta$, it follows that $\alpha\beta \leq \beta\beta$. (If $f : A \to B$ is injective, so is the map $(a, b) \to (f(a), b)$ of $A \times B$ into $B \times B$.) Thus it suffices to prove that $\beta\beta = \beta$, for then we have

$$\beta \leq \alpha\beta \leq \beta\beta = \beta.$$

We know that this holds for $\beta = \aleph_0$. Let's look at another special case, $\beta = 2^A$. (To avoid notational complications, we write 2^A rather than the formal $|2^A|$). We have

$$2^A 2^A = 2^{A+A}$$

since a pair (B, C) of subsets of A can be identified with a pair of functions from A to $\{0, 1\}$, or equivalently by an element of 2^{A+A}. But $A + A = A$ when A is infinite, so

$$2^A 2^A = 2^A$$

as desired. Now we know at least that there are very large sets satisfying $\beta\beta = \beta$.

In general, if B is a set with cardinality β, let $\mathcal{C}$ be the collection of all pairs (A, f) where $A \subseteq B$ and f is an injective map from $A \times A$ to B. Call $(A, f) \leq (A', f')$ if and only if $A \subseteq A'$ and the restriction of f' to $A \times A$ is f. Note that $\mathcal{C}$ is nonempty because B is infinite, and therefore must contain a countably infinite subset C. (Pick $b_1 \in B$; since B is infinite, there is an element $b_2 \in B$ with $b_2 \neq b_1$. Again since B is infinite, there is an element $b_3 \in B$ such that b_1, b_2, and b_3 are distinct. Continue inductively.) There must be an injective map of $C \times C$ into B because $\aleph_0 \aleph_0 = \aleph_0$. Every chain in $\mathcal{C}$ has an upper bound (A_0, f_0) where A_0 is the union of the sets A_i in the chain. If $(x, y) \in A_i \times A_i$, we take $f_0(x, y) = f_i(x, y)$, and by our definition of $\leq$ there can be no ambiguity if (x, y) also belongs to $A_j \times A_j$ for some $j \neq i$. Just as in (4.3.5), Zorn's Lemma supplies a maximal element (A^*, f^*). If $|A^*| < |B|$, select $a \in B \setminus A^*$. We have

$$(A^* \cup \{a\}) \times (A^* \cup \{a\}) = (A^* \times A^*) \cup (A^* \times \{a\}) \cup (\{a\} \times A^*) \cup (\{a\} \times \{a\}),$$

and the cardinality of the set on the right is at most $|B| + |B| + |B| + |B|$, which is $|B|$ by part (a). It follows that f^* can be extended to an injective map of $(A^* \cup \{a\}) \times (A^* \cup \{a\})$ into B, which contradicts the maximality of (A^*, f^*). Thus $|A^*| = |B|$, so there is an injective map of $B \times B$ into B, in other words, $|B||B| \leq |B|$. But we know that $|B| \leq |B||B|$ (see the beginning of the proof of part (b)), so $|B||B| = |B|$. ∎

If A is a set and n a positive integer, A^n will denote the Cartesian product $A \times A \times \cdots \times A$ (n times). If A is infinite, it follows from (4.4.3(b)) that $|A^n| = A$, $n = 1, 2, \ldots$. This implies that the collection of finite subsets of A has the same cardinality as A.

4.4.4 Corollary. *If A is infinite and $F(A)$ is the set of all finite subsets of A, then $|F(A)| = |A|$.*

Proof. The map $a \to \{a\}$ is injective, so $|A| \leq |F(A)|$. Now an element of $F(A)$ is specified by selecting a positive integer n and then a subset of A of size n. (We may safely ignore the empty set without changing the cardinality of $F(A)$.) But a subset of size n determines a point of A^n. For example, if A is the set of positive integers and $n = 3$, then $\{6, 4, 9\}$ determines the ordered triple $(4, 6, 9)$, formed by writing the elements of the set in increasing order. Since $|A^n| = |A|$, we have

$$|F(A)| \leq \aleph_0 |A| = |A|$$

by (4.4.3(b)). ∎

The discussion in (4.1.3) suggests that sets are formed by stringing together copies of the natural numbers. If we have a collection of sets, then it is plausible that we can check each set to see how far it extends along the string, and thereby identify a set in the collection of smallest size. In formal set theory, it is established that $\leq$ on cardinal numbers is not only a total ordering (see (4.3.5)), but a *well-ordering*.

Problems For Section 4.4

If A has cardinality α and B has cardinality β, we define *cardinal exponentiation* by $\alpha^\beta = |A^B|$, the cardinality of the set of functions from B to A.

1. Show that $\alpha^{\beta + \gamma} = \alpha^\beta \alpha^\gamma$.

2. Show that $(\alpha\beta)^\gamma = \alpha^\gamma \beta^\gamma$.

3. Show that $(\alpha^\beta)^\gamma = \alpha^{\beta\gamma}$.

4. Show that $\aleph_0$ is the smallest infinite cardinal.

5. Let c be the cardinality of the set of all real numbers (the "continuum"). Show that c is the same as the cardinal number of the set of reals between 0 and 1.

6. In (2.4.2) we proved that there are uncountably many subsets of the positive integers. Explicitly, the collection of subsets of the positive integers may be identified with the union of the set of real numbers between 0 and 1 and the set A of finite sequences of 0's and 1's that end in 1. (The example we gave in (2.4.2) was $.01001 = .010001111\ldots$; let's agree to use the sequence ending in all 1's to specify a real number. Then the finite sequence 01001 goes into A.) Show that A is countably infinite.

7. Show that $2^{\aleph_0} = c$.

8. The smallest uncountable cardinal, that is, the smallest cardinal greater than $\aleph_0$, is denoted by $\aleph_1$. Show that $2^{\aleph_0} \geq \aleph_1$. (The statement that $2^{\aleph_0} = \aleph_1$, and therefore $c = \aleph_1$, is called the *Continuum Hypothesis*. It cannot be proved or disproved from the axioms of set theory, even if the axiom of choice is available.)

5

Linear Algebra

5.1 Matrices

Suppose we have m linear equations expressing variables $y_1, \ldots, y_m$ in terms of variables $x_1, \ldots, x_n$:

$$y_1 = a_{11}x_1 + a_{12}x_2 + \cdots + a_{1n}x_n$$

$$y_2 = a_{21}x_1 + a_{22}x_2 + \cdots + a_{2n}x_n$$

$$\vdots$$

$$y_m = a_{m1}x_1 + a_{m2}x_2 + \cdots + a_{mn}x_n.$$

We can write these equations in the very compact form $y = Ax$, where A is an $m \times n$ matrix, that is, a rectangular array with m rows and n columns, and x and y are column vectors (x is an $n \times 1$ matrix and y is an $m \times 1$ matrix). Explicitly, using square brackets to enclose the elements of a matrix,

$$A = \begin{bmatrix} a_{11} & a_{12} & \cdots & a_{1n} \\ a_{21} & a_{22} & \cdots & a_{2n} \\ \vdots & & & \\ a_{m1} & a_{m2} & \cdots & a_{mn} \end{bmatrix}, \qquad x = \begin{bmatrix} x_1 \\ x_2 \\ \vdots \\ x_n \end{bmatrix}, \qquad y = \begin{bmatrix} y_1 \\ y_2 \\ \vdots \\ y_m \end{bmatrix}.$$

The equation $y = Ax$ makes sense if we agree that the product of the $m \times n$ matrix A and the $n \times 1$ matrix x is an $m \times 1$ matrix whose ith entry $(Ax)_i$ is found by walking across row i of A and down the column of x, multiplying terms and adding the results. Thus

$$(Ax)_i = a_{i1}x_1 + a_{i2}x_2 + \cdots + a_{in}x_n,$$

which is y_i. The advantage of matrix notation can be seen very quickly if the x_j happen to be linear combinations of variables $z_1, \ldots, z_p$. Then $x = Bz$ where B is $n \times p$ and z is $p \times 1$, and we can express y in terms of z by $y = Ax = ABz$, provided we define multiplication of A and B properly.

5.1.1 Definitions and Comments. An $m \times n$ *matrix* A is a rectangular array with m rows and n columns. If $n = 1$, we have a *column vector,* and if $m = 1$ we have a *row vector.* The ij *element* of A is the element in row i, column j of A. It is denoted by A_{ij} or a_{ij}.

To avoid an overdose of abstraction at this point, think of the components of a matrix as real numbers. In general, however, the components can belong to an arbitrary field F.

Matrices of the same size can be added componentwise. Thus if A and B are $m \times n$, then $A + B$ is the matrix whose ij element is $(A + B)_{ij} = a_{ij} + b_{ij}$; similarly, $(A - B)_{ij} = a_{ij} - b_{ij}$.

If c is a *scalar* (a member of F), then the product cA is formed by multiplying every element of A by c; thus $(cA)_{ij} = ca_{ij}$.

If A is $m \times n$ and B is $n \times p$, so that the number of columns of A is the same as the number of rows of B, then A and B can be multiplied (and we sometimes say that A and B are *compatible*). The product of A and B is an $m \times p$ matrix given by

$$(AB)_{ij} = \sum_{k=1}^{n} a_{ik} b_{kj}.$$

Thus to get the ij element of AB, we walk across row i of A and down column j of B, multiplying as we go and adding the results.

Many of the standard rules of arithmetic carry over to matrices. If $M_{mn}(F)$ is the set of all $m \times n$ matrices over F (that is, with components in the field F), then, going down the list at the beginning of Section 3.3, we have

(A1)   If A and B belong to $M_{mn}(F)$, then $A + B$ is also in $M_{mn}(F)$.

(A2) $A + (B + C) = (A + B) + C$ for all A, B, C in $M_{mn}(F)$.

(A3)   If the $m \times n$ matrix with all zeros is denoted simply by 0, then $A + 0 = 0 + A = A$ for all A in $M_{mn}(F)$.

(A4)   If $-A$ is the matrix $(-1)A$ (formed from A by multiplying every element by -1), then $A + (-A) = (-A) + A = 0$.

(A5) $A + B = B + A$ for all A, B in $M_{mn}(F)$.

(M1)   If $A \in M_{mn}(F)$ and $B \in M_{np}(F)$ then $AB \in M_{mp}(F)$.

(M2)   If $A \in M_{mn}(F)$, $B \in M_{np}(F)$ and $C \in M_{pq}(F)$, then $A(BC) = (AB)C$.

(M3)   Let I_n be the $n \times n$ *identity matrix,* that is, the $n \times n$ matrix with 1's on the main diagonal and 0's elsewhere ($a_{ii} = 1$ for $i = 1, \ldots, n$; $a_{ij} = 0$ for $i \neq j$). If $A \in M_{mn}(F)$, then $AI_n = I_m A = A$.

(M4)   If $A \in M_{mn}(F)$, $B \in M_{np}(F)$, and $C \in M_{np}(F)$, then $A(B + C) = AB + AC$; if $A \in M_{mn}(F)$, $B \in M_{mn}(F)$, and $C \in M_{np}(F)$, then $(A + B)C = AC + BC$.

These results follow from the rules of arithmetic for real numbers, or more generally for elements in any field. (See Problems 1 and 2 for some typical arguments.)

Properties (A1)–(A5) show that $M_{mn}(F)$ is an abelian group under addition. But if we attempt to prove that $M_{mn}(F)$ is a ring, we run into trouble at (M1). The product of two $m \times n$ matrices is not defined unless $m = n$. In this case we have *square matrices* with the same number of rows as columns, and we replace the notation $M_{nn}(F)$ by $M_n(F)$. Then (M1)–(M4) show that $M_n(F)$ is a ring under matrix multiplication; the multiplicative identity is the identity matrix I_n.

Even for square matrices, properties (M5),(M6) and (M7) all fail. To see this, take

$$A = \begin{bmatrix} 0 & 1 \\ 0 & 0 \end{bmatrix} \qquad \text{and} \qquad B = \begin{bmatrix} 1 & 0 \\ 0 & 0 \end{bmatrix}.$$

Then

$$AB = \begin{bmatrix} 0 & 0 \\ 0 & 0 \end{bmatrix} = 0, \qquad BA = \begin{bmatrix} 0 & 1 \\ 0 & 0 \end{bmatrix} = A$$

so that $AB \neq BA$ and in addition A and B are both zero-divisors. If A has a multiplicative inverse A^{-1}, multiplication of the equation $AB = 0$ on the left by A^{-1} yields $B = 0$, a contradiction. Similarly, if B has a multiplicative inverse B^{-1}, multiplication of the same equation on the right by B^{-1} gives $A = 0$, a contradiction.

5.1.2 Elementary Row and Column Operations.

Suppose we take a 3×3 identity matrix I_3 and interchange rows 2 and 3 to get

$$E = E(R_2 \leftrightarrow R_3) = \begin{bmatrix} 1 & 0 & 0 \\ 0 & 0 & 1 \\ 0 & 1 & 0 \end{bmatrix}.$$

If we form the product $E(R_2 \leftrightarrow R_3)A$ where A is any matrix with 3 rows and n columns, we get A with rows 2 and 3 interchanged. (Try an example, or observe that as you walk across row 2 of E and down a column of A, you pick up row 3 of A; similarly, walking across row 3 of E and down a column of A will pick up row 2 of A.) Notice that we can recover the original matrix I_3 by interchanging rows 2 and 3 again.

If in I_3 we add 5 times row 3 to row 1, we get

$$E = E(R_1 \leftarrow R_1 + 5R_3) = \begin{bmatrix} 1 & 0 & 5 \\ 0 & 1 & 0 \\ 0 & 0 & 1 \end{bmatrix},$$

and if A is $3 \times n$, the product EA is A with 5 times row 3 added to row 1. We can recover I_3 by adding -5 times row 3 of E to row 1.

If in I_3 we multiply row 2 by 13, we get

$$E = E(R_2 \leftarrow 13R_2) = \begin{bmatrix} 1 & 0 & 0 \\ 0 & 13 & 0 \\ 0 & 0 & 1 \end{bmatrix},$$

and if A is 3 by n, the product EA is A with row 2 multiplied by 13. We can recover I_3 by multiplying row 2 by $1/13$.

Operations of the form $R_i \leftrightarrow R_j$, $R_i \leftarrow R_i + kR_j$, and $R_i \leftarrow cR_i$ (where $c \neq 0$) are called *elementary row operations,* and the corresponding matrices E are called *elementary row matrices.* If E and A are compatible, then we can *premultiply* A by E (that is, form the product EA with the elementary row matrix E on the left) to get A with the row operation represented by E carried out.

In a similar fashion we have *elementary column operations* $C_i \leftrightarrow C_j$, $C_i \leftarrow C_i + kC_j$, and $C_i \leftarrow cC_i$ ($c \neq 0$), and corresponding matrices E called *elementary column matrices.* If B and E are compatible, then we can *postmultiply* B by E (that is, form the product BE with the elementary column matrix E on the right) to get A with the column operation represented by E carried out.

5.1.3 Echelon Form. Linear equations in matrix form $Ax = b$ (where b is a column vector) can be solved efficiently by reducing the matrix A, with the column b adjoined on the right, to *echelon form,* which we now describe. For example, consider

$$2x_1 + 5x_2 + 5x_3 = 1$$
$$4x_1 + 10x_2 + 0x_3 = 2$$
$$x_1 + 10x_2 + 0x_3 = 8.$$

The coefficient matrix with the right-hand column adjoined is

$$\begin{bmatrix} 2 & 5 & 5 & 1 \\ 4 & 10 & 0 & 2 \\ 1 & 10 & 0 & 8 \end{bmatrix}.$$

Interchange rows 1 and 3 to get a 1 in the 1-1 position. (Another method is to multiply row 1 by $1/2$, but the arithmetic is a bit messier that way.) The result is

$$\begin{bmatrix} 1 & 10 & 0 & 8 \\ 4 & 10 & 0 & 2 \\ 2 & 5 & 5 & 1 \end{bmatrix}.$$

Add -4 times row 1 to row 2, and then -2 times row 1 to row 3, to produce zeros in column 1 below the "pivot" in the 1-1 position; the new matrix is

$$\begin{bmatrix} 1 & 10 & 0 & 8 \\ 0 & -30 & 0 & -30 \\ 0 & -15 & 5 & -15 \end{bmatrix}.$$

Multiply row 2 by $-1/30$ to produce

$$\begin{bmatrix} 1 & 10 & 0 & 8 \\ 0 & 1 & 0 & 1 \\ 0 & -15 & 5 & -15 \end{bmatrix}.$$

Now, using the 2-2 entry as a pivot, add -10 times row 2 to row 1, and then 15 times row 2 to row 3, to get

$$\begin{bmatrix} 1 & 0 & 0 & -2 \\ 0 & 1 & 0 & 1 \\ 0 & 0 & 5 & 0 \end{bmatrix}.$$

Finally, multiply row 3 by $1/5$ to obtain the echelon form

$$\begin{bmatrix} 1 & 0 & 0 & -2 \\ 0 & 1 & 0 & 1 \\ 0 & 0 & 1 & 0 \end{bmatrix}.$$

Since the operations of interchanging equations, multiplying an equation by a constant and then adding it to another equation, or multiplying an equation by a nonzero constant, do not change the solutions to the equations, we can now read off the solution to the original problem: $x_1 = -2$, $x_2 = 1$, $x_3 = 0$.

Echelon form can be slightly more complicated. For example, consider

$$\begin{bmatrix} 1 & 3 & 0 & -12 & 0 \\ 0 & 0 & 1 & 3 & 0 \\ 0 & 0 & 0 & 0 & 1 \end{bmatrix}.$$

Columns 1, 3, and 5 are *pivot columns* containing a single 1 with all other entries 0. Non-pivot columns to the right of a pivot column have 0's below the row in which the pivot occurs. Thus column 4 has nonzero entries in rows 1 and 2, but 0 in row 3. If we try to simplify column 4, e.g., by adding 4 times row 2 to row 1, we mess up the pivot column 3. The idea is to manipulate the original matrix so as to be as close as possible to an identity matrix. Even if there are some non-pivot columns, the solution to the equations can still be read off by inspection. In the above example, the equations are inconsistent (that is, they have no solution). For although the first two equations have the benign form $x_1 + 3x_2 - 12x_4 = 0$ and $x_3 + 3x_4 = 0$, and can be satisfied by specifying x_2 and x_4 arbitrarily and solving for the pivot variables x_1 and x_3, the third equation is $0 = 1$.

If an $n \times n$ matrix A is reduced to echelon form by elementary row operations represented by elementary row matrices $E_1, \ldots, E_k$ in that order, then the echelon matrix is $E_k \cdots E_1 A$. If the echelon matrix turns out to be I_n, then the matrix $B = E_k \cdots E_1$ is the multiplicative inverse of A (see Problems 4 and 5).

5.1.4 Transpose of a Matrix. If we interchange the rows and columns of an $m \times n$ matrix A, we get an $n \times m$ matrix A^t called the *transpose* of A. For example,

$$\begin{bmatrix} 1 & 2 & 3 \\ 4 & 5 & 6 \end{bmatrix}^t = \begin{bmatrix} 1 & 4 \\ 2 & 5 \\ 3 & 6 \end{bmatrix}.$$

It follows from the definition that $(A^t)^t = A$, $(cA)^t = cA^t$, and $(A + B)^t = A^t + B^t$, assuming that A and B are of the same size. If A and B are compatible, we have

$$(AB)^t = B^t A^t,$$

the product of transposes in reverse order. To see this, note that the ij element of $(AB)^t$ is the ji element of AB, which is

$$\sum_k a_{jk} b_{ki} = \sum_k (B^t)_{ik}(A^t)_{kj},$$

the ij element of $B^t A^t$.

If A is a square matrix, then A is said to be *symmetric* if $A = A^t$ and *skew-symmetric* if $A = -A^t$. A real skew-symmetric matrix must have zeros on the main diagonal (that is, $a_{11} = \cdots = a_{nn} = 0$). This holds because the operation of transposing has no effect on the main diagonal, and therefore $a_{ii} = -a_{ii}$, which forces a_{ii} to be 0.

Since $(A + B)^t = A^t + B^t$, it follows that the sum of symmetric matrices is symmetric. But the product of symmetric matrices need not be symmetric. The following result gives precise information about this question.

5.1.5 Theorem. *Let A and B be symmetric $n \times n$ matrices. Then AB is symmetric if and only if $AB = BA$.*

Proof. If AB is symmetric then $AB = (AB)^t = B^t A^t = BA$. Conversely, assume that $AB = BA$. Then $(AB)^t = (BA)^t = A^t B^t = AB$, so that AB is symmetric. ∎

Problems For Section 5.1

1. Show that matrix multiplication obeys the distributive laws $A(B+C) = AB + AC$ and $(A+B)C = AC + BC$ (assuming the matrices are compatible).

2. Show that matrix multiplication is associative, that is, $A(BC) = (AB)C$ (assuming the matrices are compatible).

3. If $AB = AC$, does it follow that $B = C$?

4. Suppose that A is an $n \times n$ matrix, and we perform a sequence of elementary row operations on A, represented by the elementary row matrices $E_1, E_2, \ldots, E_k$ in turn. Assume that we succeed in reducing A to the identity matrix I_n (so the echelon form is as nice as it can be). If the same operations are applied to I_n, show that we obtain $B = E_k E_{k-1} \cdots E_1$. Show also that $BA = I_n$.

5. Continuing Problem 4, if $BA = I_n$ and A has a multiplicative inverse A^{-1}, show that B must be A^{-1}, so that $AB = I_n$ also. (In the next section, we will see that if $BA = I_n$, then the determinant of A (and of B) must be nonzero, and it will follow that A has an inverse, necessarily equal to B.)

6. If $A = \begin{bmatrix} 1 & 3 \\ 0 & 1 \end{bmatrix}$, find A^{74} in a relatively effortless manner.

7. If A is an arbitrary $m \times n$ matrix, show that AA^t is always symmetric.

8. We observed that if A is a real skew-symmetric matrix, then A has zeros on the main diagonal. Does this hold when the coefficients of A belong to an arbitrary field F?

9. Let A be an arbitrary real $n \times n$ matrix. Show that A can be written as the sum of a symmetric matrix and a skew-symmetric matrix.

10. Let

$$A = \begin{bmatrix} 0 & 1 & 0 & 0 \\ 0 & 0 & 1 & 0 \\ 0 & 0 & 0 & 1 \\ 0 & 0 & 0 & 0 \end{bmatrix}.$$

Show that $A^2 \neq 0$, $A^3 \neq 0$, but $A^4 = 0$. A matrix such that $A^n = 0$ for some positive integer n is said to be *nilpotent.*

5.2 Determinants and Inverses

A precise treatment of determinants would take a long time and involve intricate and unpleasant proofs. Instead, we are going to cheat a bit and assume some major properties as axioms, ignoring the question of whether there is any mathematical object at all that satisfies these axioms. This will allow us to proceed quickly to typical applications of determinants in linear algebra.

   If A is an n by n matrix with entries in the field F, the *determinant* of A (notation *det A*) is a number, in other words, an element of F.

If a matrix is given explicitly, for example,

$$A = \begin{bmatrix} 1 & 2 & 3 \\ 4 & 5 & 6 \\ 7 & 8 & 9 \end{bmatrix},$$

the determinant of A is often written with absolute value signs replacing the square brackets:

$$\det A = \begin{vmatrix} 1 & 2 & 3 \\ 4 & 5 & 6 \\ 7 & 8 & 9 \end{vmatrix}.$$

In Section 5.1 we discussed elementary row operations and the corresponding elementary row matrices. The three types of elementary row matrices will be denoted by E_{int} (corresponding to the interchange of two rows), E_{row} (corresponding to the addition of a multiple of a row to another row), and E_c (corresponding to the multiplication of a row by a nonzero constant c). The analogous elementary column matrices will be denoted by C_{int} ($= E_{\text{int}}$), C_{col} and C_c ($= E_c$). (Recall from (5.1.2) that an elementary row or column matrix can be obtained by performing the associated row or column operation on an identity matrix.) We assume

(1) If I_n is the n by n identity matrix, then $\det I_n = 1$.
(2) $\det E_{\text{int}} = \det C_{\text{int}} = -1$.
(3) $\det E_{\text{row}} = \det C_{\text{col}} = 1$.
(4) $\det E_c = \det C_c = c$.
(5) If A and B are n by n matrices, then $\det(AB) = (\det A)(\det B)$.
(6) If A has a row or column consisting entirely of 0's, then $\det A = 0$.

It follows from (2) and (5) that $\det(E_{\text{int}} A) = -\det A$; in other words, interchanging two rows of A multiplies the determinant by -1. Similarly, it follows from (3) and (5) that adding a multiple of a row to another row leaves the determinant unchanged. From (4) and (5) we see that multiplication of a row of A by c multiplies the determinant of A by c. (Analogous results hold for columns.) Since c is assumed nonzero, it follows that if E is an elementary row matrix, then $\det EA$ is nonzero if and only if $\det A$ is nonzero, and similarly for elementary column matrices.

Now suppose that A is reduced to echelon form via elementary row operations. If the echelon matrix is Q, then by the above remarks, $\det A$ and $\det Q$ are both zero or both nonzero. If Q is the identity matrix I_n, then by (1), $\det A \neq 0$. On the other hand, if Q is not I_n, then it must have a row of 0's, so that $\det Q = 0$ by (6), and consequently $\det A = 0$. Thus

(7) The echelon form of A is I_n if and only if $\det A \neq 0$.

This is the key step in the proof of the following result.

5.2.1 Theorem. *The n by n matrix A has an inverse (that is, a matrix B such that $AB = BA = I_n$) if and only if $\det A \neq 0$.*

Proof. If $AB = I_n$, then by (1) and (5), $(\det A)(\det B) = \det I_n = 1$, so $\det A \neq 0$. Thus assume $\det A \neq 0$. If A is reduced to echelon form by elementary row matrices

$E_1, \ldots, E_k$, let $B = E_k \cdots E_1$. Then by (7), $BA = I_n$, and similarly there is a product C of elementary column matrices such that $AC = I_n$. But if we multiply $BA = I_n$ on the right by C, we get $B(AC) = I_nC = C$, which says that $BI_n = C$, that is, $B = C$. Therefore $AB = BA = I_n$. ∎

Note that if $E_1, \ldots, E_k$ reduce A to I_n, the same operations applied to I_n produce $E_k \cdots E_1 I_n = B$. This gives a computational procedure for finding the inverse of A, if it exists. Furthermore, we can now compute the determinant of any square matrix A by reducing A to echelon form. We find a product B of elementary row matrices B such that BA is the echelon matrix Q. If Q is not I_n, then $\det A$ must be 0, by (7). If $Q = I_n$, then $\det A = \frac{1}{\det B} = (\det E_k)^{-1} \cdots (\det E_1)^{-1}$, which can be computed by (2), (3), and (4).

If we find by any method a matrix B such that $AB = I(= I_n)$, it is guaranteed that $BA = I$ as well, so that B is the inverse of A. For $\det A \neq 0$ by (1) and (5), so that A has an inverse by (5.2.1). Multiply $AB = I$ on the left by A^{-1} to conclude that $B = A^{-1}$.

The inverse of a matrix is unique, for if $AB = AC = I$, multiply on the left by A^{-1} to conclude that $B = C$.

If A has an inverse, we often say that A is *invertible* or *nonsingular.* Here are some additional properties of inverses.

5.2.2 Theorem. *Let* $A, A_1, \ldots, A_k$ *be* n *by* n *matrices. Then*

(a) If A is invertible, then so is A^{-1}, and $(A^{-1})^{-1} = A$.
(b) If A is invertible, then so is A^t, and $(A^t)^{-1} = (A^{-1})^t$.
(c) If $A_1, \ldots, A_k$ are all invertible, then so is the product $A_1 \cdots A_k$, and $(A_1 \cdots A_k)^{-1} = A_k^{-1} \cdots A_1^{-1}$. In particular, $(A^n)^{-1} = (A^{-1})^n$.
(d) If A is invertible and $c \neq 0$, then cA is invertible and $(cA)^{-1} = \frac{1}{c} A^{-1}$.
(e) If A is invertible, then $\det(A^{-1}) = \frac{1}{\det A}$.

Proof.

(a) $AA^{-1} = A^{-1}A = I$.
(b) Take the transpose of $AA^{-1} = I$ to get $(A^{-1})^t A^t = I$.
(c) $(A_1 \cdots A_k)(A_k^{-1} \cdots A^{-1}) = I$.
(d) $(cA)(\frac{1}{c}A^{-1}) = c(\frac{1}{c})AA^{-1} = I$.
(e) $1 = \det I = \det(AA^{-1}) = (\det A)(\det A^{-1})$. ∎

Here is another way to compute determinants: $\det([a]) = a$, $\det\begin{bmatrix} a & b \\ c & d \end{bmatrix} = ad - bc$, and for higher order determinants we use the *Laplace Expansion*: Walk across a row (or down a column), say row i. Multiply each entry a_{ij} by its *cofactor* $A(i,j)$, which is plus or minus the determinant formed by crossing out row i and column j. The plus or minus sign is determined by a checkerboard pattern, for example

$$\begin{vmatrix} + & - & + & - \\ - & + & - & + \\ + & - & + & - \\ - & + & - & + \end{vmatrix}.$$

Thus the sign chosen is $(-1)^{i+j}$. (The determinant formed by crossing out row i and column j (without attaching a plus or minus sign) is called the *minor* $M(i,j)$.) The determinant is given by $\sum_j a_{ij} A(i,j)$. It can be shown that the same answer is obtained, regardless of the row or column chosen.

Let's look at Laplace's expansion in detail for the 3 by 3 case. If

$$A = \begin{bmatrix} a & b & c \\ d & e & f \\ g & h & i \end{bmatrix}$$

and we expand across row 1, we get

$$\det A = a \begin{vmatrix} e & f \\ h & i \end{vmatrix} - b \begin{vmatrix} d & f \\ g & i \end{vmatrix} + c \begin{vmatrix} d & e \\ g & h \end{vmatrix}$$
$$= a(ei - fh) - b(di - fg) + c(dh - eg)$$
$$= aei + bfg + cdh - afh - bdi - ceg.$$

Each term is generated by selecting an element in the first row, then an element in the second row but in a new column, and finally an element in the third row and in a new column. The sign is determined as follows. The term cdh corresponds to choosing columns 3, 1, 2 in turn, and 312 is an *even permutation* of 123 in the sense that it takes an even number of transpositions (interchanges of adjacent digits) to transform 312 into the natural order 123: $312 \to 132 \to 123$ involves two transpositions. On the other hand, ceg corresponds to columns 3, 2, 1 in turn, and 321 is an *odd permutation* because it takes an odd number of transpositions to transform 321 into 123: $321 \to 312 \to 132 \to 123$. In general, the determinant of an n by n matrix can be defined as the sum of all $n!$ terms formed by selecting an element in row 1, then an element in row 2 but in a new column,..., and finally an element in row n but in a new column. A plus sign is attached to a term if its successive column choices yield an even permutation of $12\ldots n$, and a minus sign is attached if the column choices yield an odd permutation. The determinant is the sum of the resulting terms. We will get the same result if we choose an element in column 1, then an element in column 2 but in a new row, etc.

Laplace's expansion and the basic properties of determinants can be used to derive a familiar formula for the solution of a set of linear equations.

5.2.3 Cramer's Rule.   Let A be a nonsingular n by n matrix, and let $\operatorname{Adj} A$ (the *adjoint* of A) be the transposed matrix of cofactors, that is, the ij element of $\operatorname{Adj} A$ is $A(j,i)$. Then

$$A^{-1} = \frac{\operatorname{Adj} A}{\det A}.$$

Furthermore, if B_i is the matrix obtained by replacing column i of A by the column vector b, then the equations $Ax = b$ have the unique solution

$$x_i = \frac{\det B_i}{\det A}, \quad i = 1, \ldots, n.$$

Proof. The Laplace expansion across row i is given by

$$a_{i1}A(i,1) + a_{i2}A(i,2) + \cdots + a_{in}A(i,n) = \det A. \tag{8}$$

But the cofactors $A(i,j)$, $j = 1, \ldots, n$, form row i of the matrix of cofactors, which is column i of the transposed matrix of cofactors, namely Adj A. Thus (8) says that the ii element of A Adj A is $\det A$, so the ii element of $A\left(\frac{\text{Adj } A}{\det A}\right)$ is 1. Now we claim that if $i \ne j$, then

$$a_{i1}A(j,1) + a_{i2}A(j,2) + \cdots + a_{in}A(j,n) = 0. \tag{9}$$

To see this, replace row j of A by a copy of row i, producing a matrix C whose determinant is 0 (see Problem 3). But the Laplace expansion across row j of C is precisely the left side of (9), and since the Laplace expansion is $\det C = 0$, (9) is proved. But this equation says that the ij element of A Adj A, and therefore the ij element of $A\left(\frac{\text{Adj } A}{\det A}\right)$, is 0 for $i \ne j$.

To summarize, we have $A\left(\frac{\text{Adj } A}{\det A}\right) = I$, so $A^{-1} = \frac{\text{Adj } A}{\det A}$.

Furthermore, the equations $Ax = b$ have the unique solution $x = A^{-1}b$. Walking across row i of A^{-1} and down the column of b, we find that

$$x_i = (\det A)^{-1}\left(A(1,i)b_1 + A(2,i)b_2 + \cdots + A(n,i)b_n\right).$$

(To verify this, note that $A(1,i), \ldots, A(n,i)$ form column i of the matrix of cofactors, which is row i of Adj A.) But $A(1,i)b_1 + \cdots A(n,i)b_n$ is the Laplace expansion down column i of the matrix B_i, so we have

$$x_i = \frac{\det B_i}{\det A}. \quad \blacksquare$$

Cramer's Rule gives a simple explicit form for the inverse of a 2 by 2 matrix with nonzero determinant:

$$\begin{bmatrix} a & b \\ c & d \end{bmatrix}^{-1} = \frac{1}{ad - bc}\begin{bmatrix} d & -b \\ -c & a \end{bmatrix}.$$

5.2.4 Example. Let

$$A = \begin{bmatrix} 0 & 1 & 4 \\ 1 & 0 & 0 \\ 0 & -2 & 2 \end{bmatrix}.$$

We will find A^{-1} by elementary row operations, and also by Cramer's Rule. We display A and I side by side:

$$\begin{array}{ccc} 0 & 1 & 4 \\ 1 & 0 & 0 \\ 0 & -2 & 2 \end{array} \qquad \begin{array}{ccc} 1 & 0 & 0 \\ 0 & 1 & 0. \\ 0 & 0 & 1 \end{array}$$

We apply elementary row operations to reduce A to I, and carry out the same operations on I:

$R_1 \leftrightarrow R_2$

$$
\begin{array}{ccc}
1 & 0 & 0 \\
0 & 1 & 4 \\
0 & -2 & 2
\end{array}
\qquad
\begin{array}{ccc}
0 & 1 & 0 \\
1 & 0 & 0 \\
0 & 0 & 1
\end{array}
$$

$R_3 \leftarrow R_3 + 2R_2$

$$
\begin{array}{ccc}
1 & 0 & 0 \\
0 & 1 & 4 \\
0 & 0 & 10
\end{array}
\qquad
\begin{array}{ccc}
0 & 1 & 0 \\
1 & 0 & 0 \\
2 & 0 & 1
\end{array}
$$

$R_3 \leftarrow \frac{1}{10} R_3$

$$
\begin{array}{ccc}
1 & 0 & 0 \\
0 & 1 & 4 \\
0 & 0 & 1
\end{array}
\qquad
\begin{array}{ccc}
0 & 1 & 0 \\
1 & 0 & 0 \\
\frac{1}{5} & 0 & \frac{1}{10}
\end{array}
$$

$R_2 \leftarrow R_2 - 4R_3$

$$
\begin{array}{ccc}
1 & 0 & 0 \\
0 & 1 & 0 \\
0 & 0 & 1
\end{array}
\qquad
\begin{array}{ccc}
0 & 1 & 0 \\
\frac{1}{5} & 0 & -\frac{2}{5} \\
\frac{1}{5} & 0 & \frac{1}{10}
\end{array}
$$

Therefore

$$
A^{-1} = \begin{bmatrix} 0 & 1 & 0 \\ \frac{1}{5} & 0 & -\frac{2}{5} \\ \frac{1}{5} & 0 & \frac{1}{10} \end{bmatrix}.
$$

Notice also that the first row operation multiplies the determinant by -1, and the third operation multiplies the determinant by $1/10$; the other operations leave the determinant unchanged. Since the determinant of I is 1, it follows that $\det A = \frac{1}{-1/10} = -10$. In this case, Laplace expansion down column 1 gets the result more quickly.

To calculate the inverse by Cramer's Rule (a much less efficient method for large matrices), we compute cofactors:

$$
A(1,1) = \begin{vmatrix} 0 & 0 \\ -2 & 2 \end{vmatrix} = 0, \quad A(1,2) = -\begin{vmatrix} 1 & 0 \\ 0 & 2 \end{vmatrix} = -2, \quad A(1,3) = \begin{vmatrix} 1 & 0 \\ 0 & -2 \end{vmatrix} = -2,
$$

$$
A(2,1) = -\begin{vmatrix} 1 & 4 \\ -2 & 2 \end{vmatrix} = -10, \quad A(2,2) = \begin{vmatrix} 0 & 4 \\ 0 & 2 \end{vmatrix} = 0, \quad A(2,3) = -\begin{vmatrix} 0 & 1 \\ 0 & -2 \end{vmatrix} = 0,
$$

$$
A(3,1) = \begin{vmatrix} 1 & 4 \\ 0 & 0 \end{vmatrix} = 0, \quad A(3,2) = -\begin{vmatrix} 0 & 4 \\ 1 & 0 \end{vmatrix} = 4, \quad A(3,3) = \begin{vmatrix} 0 & 1 \\ 1 & 0 \end{vmatrix} = -1.
$$

The adjoint matrix is

$$
\text{Adj } A = \begin{bmatrix} 0 & -10 & 0 \\ -2 & 0 & 4 \\ -2 & 0 & -1 \end{bmatrix},
$$

and since $\det A = -10$, we have

$$
A^{-1} = \frac{\text{Adj } A}{\det A} = \begin{bmatrix} 0 & 1 & 0 \\ \frac{1}{5} & 0 & -\frac{2}{5} \\ \frac{1}{5} & 0 & \frac{1}{10} \end{bmatrix}
$$

as above.

Problems For Section 5.2

1. Find the inverse of

$$A = \begin{bmatrix} 1 & 3 & 0 \\ 2 & 1 & 0 \\ 0 & 1 & -1 \end{bmatrix}$$

by using elementary row operations.

2. Use the elementary row operations in your solution of Problem 1 to find the determinant of A. Check your result by Laplace Expansion on a convenient row or column.

In Problems 3, 4, and 5, A is an arbitrary $n \times n$ matrix.

3. If A has two identical rows (or columns), show that $\det A = 0$.

4. If row i of A is a linear combination of other rows of A, show that $\det A = 0$. (A similar result holds for columns.)

5. Show that $\det A^t = \det A$.

5.3 The Vector Space F^n; Linear Independence and Bases

First we will describe a space that is a natural generalization of Euclidean two-and three-space, and that provides a very convenient setting for concrete computations. If $u_1, \ldots, u_n$ are real numbers, or more generally belong to the field F, we can form the n-tuple $u = (u_1, \ldots, u_n)$, which will be called a n-dimensional *vector*. To save space, u will be written as a row vector, but if A is an $m \times n$ matrix and we write the expression Au, we assume that u has been converted into a column vector. (Otherwise Au does not make sense.) Vectors can be added, subtracted and multiplied by scalars, so the collection of all n-dimensional vectors with components in the field F is a *vector space* over F, denoted by F^n.

At the end of Section 4.2, we gave the formal definition of a vector space, and we should convince ourselves that F^n does indeed satisfy the requirements. But nothing more than the definition of a field is involved. For example, if $a \in F$ and $u, v \in F^n$, then

$$a(u + v) = a\big((u_1, \ldots, u_n) + (v_1, \ldots, v_n)\big) = a\big((u_1 + v_1, \ldots, u_n + v_n)\big)$$

$$= \big(a(u_1 + v_1), \ldots, a(u_n + v_n)\big) = (au_1 + av_1, \ldots, au_n + av_n)$$

$$= (au_1, \ldots, au_n) + (av_1, \ldots, av_n) = au + av.$$

Similar calculations justify the other properties.

If S is a set of vectors (in other words, a subset of an arbitrary vector space V), we say that S is *linearly independent* (or that the vectors in S are linearly independent) if no vector in S can be expressed as a finite linear combination of other vectors in S; otherwise, S is *linearly dependent*. These ideas are familiar from calculus. In Euclidean 3-space $\mathbb{R}^3$, with a vector represented by an arrow whose tail is at the origin, two vectors are linearly dependent if and only if they are collinear. Three vectors are linearly dependent if and only if they are coplanar. More than three vectors in $\mathbb{R}^3$ are always linearly dependent.

If the vector u is a linear combination of the vectors v and w, for example if $u = 5v - 16w$, then $u - 5v + 16w = 0$. Conversely, if $u - 5v + 16w = 0$, then each of the three vectors can be expressed as a linear combination of the other two. It follows that S is linearly independent if and only if for all $n = 1, 2, \ldots$, all $u_1, \ldots, u_n \in S$ and all $a_1, \ldots, a_n \in F$,

$$a_1 u_1 + \cdots + a_n u_n = 0 \quad \text{implies} \quad a_1 = \cdots = a_n = 0.$$

(If you have trouble with the formal details, see the solution to Problems 2 and 3 of Section 4.2.)

From what we know about Euclidean 3-space, a natural conjecture is that we can find n linearly independent vectors in F^n, but no more. A convenient linearly independent set of size n is formed by the *coordinate vectors*

$$e_1 = (1, 0, 0, \ldots, 0)$$

$$e_2 = (0, 1, 0, \ldots, 0)$$

$$\vdots$$

$$e_n = (0, 0, 0, \ldots, 1).$$

If $a_1, \ldots, a_n \in F$, then $a_1 e_1 + \cdots + a_n e_n = (a_1, \ldots, a_n)$, and this is the zero vector if and only if all $a_i = 0$. Thus $e_1, \ldots, e_n$ are linearly independent. Furthermore, if $u = (a_1, \ldots, a_n)$ is any vector in F^n, then $e_1, \ldots, e_n, u$ are linearly dependent. (We have just seen that $u = a_1 e_1 + \cdots + a_n e_n$.) So there is no way to add a vector to the set $\{e_1, \ldots, e_n\}$ without producing linear dependence.

Let's look carefully at what we know so far. If u is any vector in F^n, then $\{e_1, \ldots, e_n, u\}$ is a linearly dependent set. We have not yet shown that *every* set of $n + 1$ vectors in F^n is linearly dependent; let's try to clarify the situation.

5.3.1 Definitions. Let V be a vector space over the field F. The subset S of V is said to *span* or *generate* V if every vector in V can be expressed as a finite linear combination of elements of S. If $S = \{x_1, x_2, \ldots\}$, we also say that the vectors $x_1, x_2, \ldots$ span V. If V has a finite spanning set, we say that V is *finite-dimensional*. A *basis* for V is a subset S that is linearly independent and spans V.

The above discussion shows that $\{e_1, \ldots, e_n\}$ is a basis for F^n, called the *standard basis*.

5.3.2 Theorem. *Let V be a finite-dimensional vector space over F. If $\{x_1, \ldots, x_m\}$ spans V and S is a linearly independent subset of V, then $|S| \leq m$.*

Proof. Suppose that S has $m + 1$ elements $y_1, \ldots, y_{m+1}$. Since the x_i span V, we can write $y_1 = \sum_{i=1}^{m} a_i x_i$ for some a_i's in F. Not all a_i can be 0, for if so we have $y_1 = 0$, which means that S is linearly dependent. (For example, $b_1 y_1 = 0$ for all $b_1 \in F$.) We may relabel the a_i and x_i so that $a_1 \neq 0$, and then

$$x_1 = a_1^{-1} y_1 - \sum_{i=2}^{m} a_1^{-1} a_i x_i.$$

It follows that $\{y_1, x_2, \ldots, x_m\}$ spans V. We claim that for every $i = 1, \ldots, m$, $\{y_1, \ldots, y_i, x_{i+1}, \ldots, x_m\}$ spans V. (This technique is called the *Steinitz Exchange*; we are replacing x's by y's in the spanning set.) We have just proved the $i = 1$ case. Now, assuming that for a specific i, $\{y_1, \ldots, y_i, x_{i+1}, \ldots, x_m\}$ spans V (the induction hypothesis), we may write

$$y_{i+1} = \sum_{j=1}^{i} a_j y_j + \sum_{k=i+1}^{m} a_k x_k.$$

Not all a_k in the second summation can be 0, for this would mean that S is linearly dependent. As above, we may assume **without loss of generality** that $a_{i+1} \neq 0$. (Notice the highlighted phrase. It is a very compact device that mathematicians use in order to make convenient additional assumptions without compromising the validity of an argument.) Then

$$x_{i+1} = -\sum_{j=1}^{i} a_{i+1}^{-1} a_j y_j + a_{i+1}^{-1} y_{i+1} - \sum_{k=i+2}^{m} a_{i+1}^{-1} a_k x_k,$$

which implies that $\{y_1, \ldots, y_{i+1}, x_{i+2}, \ldots, x_m\}$ spans V, proving the claim. Since there are more y's than x's, eventually the x's disappear, leaving $y_1, \ldots, y_m$ to generate V. But then y_{m+1} is a linear combination of $y_1, \ldots, y_m$, contradicting the assumption that S is linearly independent. Thus our original assumption that S has more than m elements is untenable, and the result follows. ∎

Theorem 5.3.2 gives us very precise information about bases, linearly independent subsets and spanning sets.

5.3.3 Theorem. *Let V be a finite-dimensional vector space, and let S and T be subsets of V.*
 (a) *V has a finite basis.*
 (b) *If S and T are bases of V, then $|S| = |T|$; this number is called the dimension of V, abbreviated $\dim V$.*
 (c) *S is a basis for V if and only if S is a maximal linearly independent set. Therefore any linearly independent subset of V can be extended (that is, enlarged) to a basis.*
 (d) *Any n linearly independent elements in an n-dimensional space form a basis.*
 (e) *S is a basis for V if and only if S is a minimal spanning set, that is, no proper subset of S spans V. Therefore any spanning set contains a basis.*
 (f) *Any n vectors that span an n-dimensional space form a basis.*
 (g) *If V is n-dimensional, then every set of $n + 1$ vectors in V is linearly dependent.*

Proof.
    (a)   Let $T = \{x_1, \ldots, x_m\}$ be a finite spanning set. If T is not linearly independent, then one of the vectors is a linear combination of the others, and this vector can be removed from T and the remaining x_i will still span V. Continue removing vectors until we reach a linearly independent spanning set, in other words, a basis.

    (b)   By (5.3.2), $|S| \leq |T|$ and $|T| \leq |S|$.

    (c)   Let S be a linearly independent subset of V. (If S is not linearly independent, both sides of the "if and only if" statement are false, and there is nothing to prove.) If

$S \subset T$ with T linearly independent, let u belong to T but not to S. Then u cannot be expressed as a finite linear combination of elements of S (this would make T linearly dependent). But then S cannot span V. Conversely, if S doesn't span, then there is some vector u that cannot be expressed as a linear combination of elements of S, so adding u to S produces a larger linearly independent set. Finally, if T is a linearly independent subset of V, then if T is maximal, it is a basis. If T is not maximal, then T is a proper subset of a linearly independent set T_1. If T_1 is maximal, it is a basis, and if T_1 is not maximal, it can be enlarged to a linearly independent set T_2. Continue inductively until a basis is reached.

(d) By (5.3.2), any linearly independent subset of V has at most n elements, so any n linearly independent elements form a maximal linearly independent set, which is a basis by (c).

(e) Let S span V. (As in (c), if S doesn't span there is nothing to prove.) If a proper subset of S spans V, then S cannot be linearly independent, and therefore cannot be a basis. Conversely, if S is not linearly independent, there is an element of S expressible as a linear combination of other elements. Deleting this element produces a smaller spanning set. An inductive version of this argument shows that any spanning set contains a basis.

(f) Let S be a set of n vectors that span the n-dimensional space V. By (e), S contains a basis B, which by (b) must have n elements. But since S also has n elements, we must have $S = B$.

(g) If V has $n+1$ linearly independent vectors, then by (c), V has a basis consisting of at least $n + 1$ elements, contradicting the assumption that V is n-dimensional. ■

If V consists of the zero vector alone, we will regard V as a *zero-dimensional space* with the empty set as basis.

Suppose we wish to test the vectors $u_1, \ldots, u_m$ for linear independence. One convenient way is to line them up as columns, forming the matrix

$$A = [\, u_1 \quad \ldots \quad u_m \,].$$

If x is a column vector with entries $a_1, \ldots, a_m$, then by visualizing a walk across a row of A and down column x, we see that $Ax = a_1 u_1 + \cdots + a_m u_m$. Thus the u_i are linearly independent if and only if the equations $Ax = 0$ have only the all zero solution. In practice, the equations may be solved by reducing A to echelon form, as in Section 5.1.

5.3.4 Example. Let $u_1 = (1, 0, -2, 2)$, $u_2 = (0, 0, 1, 1)$, $u_3 = (2, 0, -1, 4)$, $u_4 = (0, 1, 0, 0)$. Are the u_i linearly dependent? If so, find an explicit linear relation among them.

The equations $au_1 + bu_2 + cu_3 + du_4 = 0$ can be written as

$$1a + 0b + 2c + 0d = 0$$

$$0a + 0b + 0c + 1d = 0$$

$$-2a + 1b - 1c + 0d = 0$$

$$2a + 1b + 4c + 0d = 0.$$

A reduction of the coefficient matrix to echelon form yields

$$\begin{bmatrix} 1 & 0 & 2 & 0 \\ 0 & 1 & 3 & 0 \\ 0 & 0 & 0 & 1 \\ 0 & 0 & 0 & 0 \end{bmatrix},$$

and consequently $a = -2c$, $b = -3c$, $d = 0$, with c arbitrary. Thus there are many possible choices of (a, b, c, d) such that $au_1 + bu_2 + cu_3 + du_4 = 0$, and the u_i are linearly dependent.

Problems For Section 5.3

1. Show that $S = \{u_1, \ldots, u_n\}$ is a basis for the vector space V if and only if each $x \in V$ can be expressed as $a_1 u_1 + \cdots + a_n u_n$, where the coefficients a_i are unique.

2. Let $u = (1, 0, 0)$, $v = (0, 1, 1)$, $w = (0, 0, 1)$. Show that u, v, and w form a basis for $\mathbb{R}^3$.

3. In Problem 2, if $x = (2, 3, 4)$, express x as a linear combination of u, v, and w.

4. Let V be a vector space that is not necessarily finite-dimensional. If S and T are bases for V, show that S and T are both finite (and hence of equal size) or both infinite.

5. Let S and T be bases for the vector space V; we are going to show that S and T have the same cardinality. In view of Problem 4, we may assume that both S and T are infinite. Let T consist of elements y_i, where i ranges over the index set I. (Thus $|T| = |I|$.) If $x \in S$, then x has a unique expression $a_1 y_{i_1} + \cdots + a_r y_{i_r}$ in terms of the y_i. Define $I(x) = \{i_1, \ldots, i_r\}$, the set of indices used in the expansion of x in terms of the basis vectors of T. Show that $I = \cup \{I(x) : x \in S\}$.

6. Continuing Problem 5, show that $|I| \leq |S| \aleph_0 \ (= |S|$ since S is infinite). Conclude that $|T| \leq |S|$, so by symmetry, $|T| = |S|$.

5.4 Subspaces

If u and v are two linearly independent vectors in Euclidean 3-space $\mathbb{R}^3$, then the vectors $au + bv$, where a and b are arbitrary real numbers, form a plane P through the origin. This plane is a two-dimensional vector space with basis $\{u, v\}$. How do we know that P is a vector space? Since P is a subset of $\mathbb{R}^3$, the elements of P will obey all the required rules of vector arithmetic. But it would not be correct to assert that any subset of $\mathbb{R}^3$ is a vector space. For example, let P' be the set of all (a, b, c) such that $a = 5$. Then $(5, 0, 1)$ and $(5, 2, 7)$ belong to P', but the sum $(10, 2, 8)$ does not. Equally well, $(5, 0, 1) \in P'$ but $3(5, 0, 1) = (15, 0, 3) \notin P'$. The key requirement is that we must be able to add (or subtract) vectors or multiply vectors by scalars and stay within the space. In the case of P, there is no problem. For example, $(3u - 6v) + (12u + 4v) = 15u - 2v \in P$; $7(3u - 6v) = 21u - 42v \in P$.

5.4.1 Definitions and Comments. A *subspace* W of the vector space V is a subset of V that is also a vector space. Equivalently, W is a subset of V that is *closed under addition and scalar multiplication.* In other words:

> if u and v belong to W then $u + v$ is in W; and
> if $u \in W$ and $a \in F$ then $au \in W$.

These two conditions can be merged into one:

> if $u, v \in W$ and $a, b \in F$, then $au + bv \in W$.

We say that W is *closed under linear combination.*

If $u_1, \ldots, u_k$ are arbitrary vectors in V, the *subspace* (or simply the *space*) *spanned by the u_i* is the set of all vectors $a_1 u_1 + \cdots + a_k u_k, a_1, \ldots, a_k \in F$.

There is an efficient procedure for finding a basis for the subspace spanned by a given set of vectors. For example, suppose we are given four vectors u, v, w, p in F^4. To find a basis for the space spanned by these vectors, line them up as rows of a matrix A and row-operate A into echelon form. Suppose the echelon form looks like this:

$$B = \begin{bmatrix} 1 & 0 & 0 & 3 \\ 0 & 1 & 0 & -1.5 \\ 0 & 0 & 1 & 6 \\ 0 & 0 & 0 & 0 \end{bmatrix}.$$

An elementary row operation replaces row R_i by a nonzero multiple of R_i or by another row R_j or by a linear combination of R_i and R_j. Thus:

Elementary row operations do not change the row space, that is, the space spanned by the rows of the matrix. It follows that the first three rows of B (the nonzero echelon rows) form a basis for the space spanned by u, v, w, and p.

Another approach is to extract a basis from among the original vectors u, v, w, p; this is always possible by (5.3.3(e)). In this case we line up u, v, w, p as *columns* of a matrix C, and *row-operate C into echelon form.*

Elementary row operations do change the column space of C, but *elementary row operations do not change the dependence relations among the columns.* For example, if C_i is column i of C, suppose that $C_3 = 5C_1 - 12C_2$, and we add -4 times row 1 to row 3. Then the third entry of row 1 is 5 times the first entry minus 12 times the second entry, a relationship that still holds after row 1 is multiplied by -4, and which holds as well in row 3, before and after we add -4 times row 1 to row 3. If, for example, it turns out that the echelon form is

$$\begin{bmatrix} 1 & 0 & 0 & 2 \\ 0 & 1 & 0 & 6 \\ 0 & 0 & 1 & -3 \\ 0 & 0 & 0 & 0 \end{bmatrix},$$

then we know that columns 1, 2, and 3 (in the original matrix C as well as in the echelon form) are linearly independent. Thus u, v, and w form a basis for the subspace.

Here is an elegant way of showing that elementary row operations do not change dependence relations among the columns. If a matrix $[c_1, \ldots, c_n]$ with columns $c_1, \ldots, c_n$ is premultiplied by an elementary row matrix E, the result is the matrix $[Ec_1, \ldots, Ec_n]$.

If $a_1, \ldots, a_n$ are arbitrary elements in the field F, then

$$a_1 c_1 + \cdots + a_n c_n = 0$$

if and only if

$$E(a_1 c_1 + \cdots + a_n c_n) = 0$$

[since E is invertible (see (5.1.2))] if and only if

$$a_1 E(c_1) + \cdots + a_n E(c_n) = 0$$

[by definition of matrix addition and multiplication].

To determine whether a vector y belongs to the subspace spanned by a given set of vectors, say u, v, w, again line up u, v, w as columns to form a matrix A. Then y can be expressed as $au + bv + cw$ if and only if $Ax = y$, where x is the column vector with components a, b, c. So the question comes down to whether the equation $Ax = y$ has a solution, and this can be determined by reducing to echelon form, as in (5.1.3).

Here is another application of the reduction of a matrix to echelon form.

5.4.2 Theorem.   *For a given matrix A, the following numbers are all equal:*

(a) *The maximal number of linearly independent rows (i.e., the dimension of the row space of A).*

(b) *The maximal number of linearly independent columns (i.e., the dimension of the column space of A).*

(c) *The number of nonzero echelon rows.*

(d) *The number of echelon columns with pivots.*

(e) *The order of the largest nonvanishing subdeterminant of A.*

The unique number given by (a)−(e) *is called the* **rank** *of A.*

Proof.   (a) and (c) are equal because elementary row operations do not change the row space, and (b) and (d) are equal because elementary row operations do not change the dependence relations among the columns. You can convince yourself that (c) = (d) by staring at a typical echelon form (see (5.1.3)). To show that (e) belongs on the list, note that a particular set of k rows and k columns of A will combine to produce a nonvanishing subdeterminant if and only if elementary row operations can produce an identity matrix of size $k \times k$ in these rows and columns (see (7) of Section 5.2). Thus the existence of a nonvanishing subdeterminant formed from a particular set of k columns is equivalent to linear independence of the given columns, since elementary row operations do not change column dependence relations. (An echelon matrix other than I_k will contain non-pivot columns, which will be linearly dependent on previous pivot columns.) ∎

We may now give a sizable number of conditions that are equivalent to invertibility of a matrix.

5.4.3 Theorem.   *Let A be an n by n matrix. The following conditions are* equivalent, *in other words, the conditions are either all true or all false.*

(1) Λ *is invertible (nonsingular).*
(2) $\det A \neq 0$.
(3) *The echelon form of A is I_n.*
(4) *The rows of A are linearly independent.*
(5) *The columns of A are linearly independent.*
(6) *The rank of A is n.*

Proof. (1), (2), and (3) are equivalent by (5.2.1) and (7) of Section 5.2; (4), (5), and (6) are equivalent by the "(a) = (b)" part of (5.4.2). By (e) of (5.4.2), (2) is equivalent to (6). ∎

Subspaces appear in a natural and significant way when we solve *homogeneous linear equations* $Ax = 0$, where A is an m by n matrix, x is a column vector with n components, and 0 stands for a column of m zeros. Suppose, for example, that A row-operates to the echelon form

$$\begin{bmatrix} 1 & 3 & 0 & -2 & 4 & 0 & -1 \\ 0 & 0 & 1 & 2 & 6 & 0 & -2 \\ 0 & 0 & 0 & 0 & 0 & 1 & 0 \\ 0 & 0 & 0 & 0 & 0 & 0 & 0 \end{bmatrix}.$$

We can if we like adjoin a column of zeros on the right, but this will never be changed by the elementary row operations, so we may as well omit it. The equations are:

$$x_1 + 3x_2 - 2x_4 + 4x_5 - x_7 = 0$$

$$x_3 + 2x_4 + 6x_5 - 2x_7 = 0$$

$$x_6 = 0.$$

Thus we may specify x_2, x_4, x_5, and x_7 (the variables corresponding to non-pivot columns) arbitrarily, and solve for x_1, x_3, and x_6. If $x_2 = a$, $x_4 = b$, $x_5 = c$, and $x_7 = d$, we have

$$x_1 = -3a + 2b - 4c + d$$

$$x_2 = a$$

$$x_3 = -2b - 6c + 2d$$

$$x_4 = b$$

$$x_5 = c$$

$$x_6 = 0$$

$$x_7 = d.$$

In vector notation,

$$x = (x_1, x_2, x_3, x_4, x_5, x_6, x_7) = a(-3, 1, 0, 0, 0, 0, 0) + b(2, 0, -2, 1, 0, 0, 0)$$

$$+ c(-4, 0, -6, 0, 1, 0, 0) + d(1, 0, 2, 0, 0, 0, 1)$$

$$= au + bv + cw + dp.$$

Note that u corresponds to the solution with $a = 1$, $b = c = d = 0$; similarly, v corresponds to $a = 0$, $b = 1$, $c = 0$, $d = 0$, w to $a = b = 0$, $c = 1$, $d = 0$, and p to $a = b = c = 0$, $d = 1$.

The set of solutions coincides with the set of linear combinations of u, v, w, and p, in other words, the subspace spanned by u, v, w, and p. Furthermore, u, v, w, and p are linearly independent, and therefore form a basis for the subspace. (Look at the second, fourth, fifth, and seventh components to see that none of the vectors can be expressed as a linear combination of the others.)

Thus the variables x_2, x_4, x_5, and x_7 that are specified arbitrarily determine a basis u, v, w, p for the solution space. The remaining three variables x_1, x_3, and x_6 correspond to pivot columns of A, and thus the rank of A is 3.

This example is typical of the general case—

5.4.4 Homogeneous Linear Equations. If A is an m by n matrix of rank r, then the set of solutions of $Ax = 0$ is called the *null space* of A, written $N(A)$. The null space is an $(n - r)$-dimensional subspace of F^n, and the dimension $n - r$ is sometimes called the *nullity* of A. If A is reduced to echelon form by elementary row operations, then the $n - r$ variables corresponding to non-pivot columns can be specified arbitrarily. The r variables corresponding to pivot columns are then determined.

Looking again at the above example, let's examine the column space of A in more detail. If the columns of A are $u_1, \ldots, u_n$ and x is a column vector with components $a_1, \ldots, a_n$, then $Ax = a_1 u_1 + \cdots + a_n u_n$ (see the discussion preceding Example 5.3.4). Thus the column space of A coincides with set of vectors Ax where x ranges over all of F^n. This set is called the *range* of A, written $R(A)$. Since the columns of A which yield echelon columns with pivots form a basis for the column space (see (b) and (d) of (5.4.2)), the range of A has dimension r. We therefore have the following fundamental result:

5.4.5 Dimension Theorem. *If A is an m by n matrix of rank r, then* $\dim N(A) = n - r$, $\dim R(A) = r$; *consequently* $\dim N(A) + \dim R(A) = n$. *In fact if $u_1, \ldots, u_k$, $k = n - r$, form a basis for $N(A)$, and the u_i are extended to a basis $u_1, \ldots, u_n$ for F^n, then $Au_{k+1}, \ldots, Au_n$ form a basis for $R(A)$.*

Proof. If $Ax \in R(A)$, express x in terms of the u_i as $a_1 u_1 + \cdots + a_n u_n$. Since u_i belongs to $N(A)$ for $i = 1, \ldots, k$, we have $Au_i = 0$, so that $Ax = \sum_{i=k+1}^{n} a_i Au_i$. It follows that $Au_{k+1}, \ldots, Au_n$ span $R(A)$. If $\sum_{i=k+1}^{n} b_i Au_i = 0$, then $\sum_{i=k+1}^{n} b_i u_i$ belongs to the null space $N(A)$, so that it can be expressed as a linear combination of $u_1, \ldots, u_k$. But $\{u_1, \ldots, u_n\}$ is a basis for F^n, so we must have $b_{k+1} = \cdots = b_n = 0$. Thus the Au_i, $k + 1 \leq i \leq n$, are linearly independent, completing the proof. ∎

5.4.6 Nonhomogeneous Linear Equations. The set of solutions to $Ax = b$ is not a subspace when $b \neq 0$; for example, if $Ax = b$ then $A(2x) = 2b \neq b$. But in a sense, the set of solutions comes close to being a subspace. If x_0 is a particular solution, then for any solution x, $x - x_0$ satisfies the homogeneous system $Ay = 0$, since $A(x - x_0) = Ax - Ax_0 = b - b = 0$. Conversely, if $Ay = 0$, then $A(x_0 + y) = Ax_0 + Ay = b + 0 = b$.

Thus if S is the subspace of homogeneous solutions, we can generate all nonhomogeneous solutions by translating S by x_0. For short,

$$\text{nonhomogeous solutions} \ = \ \text{homogeneous solutions} \ + \ \text{particular solution}$$

In the homogeneous case, there is always at least one solution (namely 0), but nonhomogeneous equations can be inconsistent. (See the end of (5.1.3) for an example.) Thus the particular solution x_0 might not exist. A systematic method for solving linear equations such as reduction to echelon form will detect inconsistent equations.

Problems For Section 5.4

1. Find a subset of $\mathbb{R}^2$ that is
   (a) closed under addition but not closed under scalar multiplication;
   (b) closed under scalar multiplication but not closed under addition.

2. Let S be the set of all vectors in $\mathbb{R}^4$ of the form $(a, b, c, 2a - b)$, where a, b, and c are arbitrary real numbers. Show that S is a subspace.

3. In Problem 2, find a basis for S.

4. If u, v, and w are a basis for a three-dimensional subspace S, show that $u+v$, $v+w$, and $w + u$ also form a basis for S.

5. Show that the vectors $u = (1, 0, 2)$, $v = (2, 1, 3)$, and $w = (0, 1, 1)$ form a basis for $\mathbb{R}^3$, and express $(1, 4, 8)$ in terms of this basis.

6. Let A be an m by n matrix with null space $K = N(A)$ and range $R(A)$. For a fixed $u \in F^n$, let C be the set of all vectors $u+v$, $v \in K$. (C is called a *coset of K* or a *coset modulo K*, and is written as $u + K$; this idea is critically important in abstract algebra and algebraic geometry.)
   (a) Give a geometric interpretation of C in Euclidean 2- or 3-space.
   (b) Let π be the mapping that assigns to the coset $u + K$ the element Au. Show that π is well-defined; that is, if $u + K = v + K$, then $Au = Av$.

7. Continuing Problem 6, define the sum of two cosets $u+K$ and $v+K$ as $(u+v)+K$, and define scalar multiplication of a coset by $a(u + K) = au + K$.
   (a) Show that addition and scalar multiplication are well-defined; that is, if $u_1+K = u_2+K$ and $v_1+K = v_2+K$, then $(u_1+v_1)+K = (u_2+v_2)+K$ and $au_1+K = au_2+K$.
   (b) Show that the mapping π is linear; that is,

$$\pi\big(a(u + K) + b(v + K)\big) = a\pi(u + K) + b\pi(v + K).$$

   (c) Show that π is bijective.
   If V is the vector space F^n and K is the subspace $N(A)$, the results of Problems 6 and 7 show that the collection of cosets modulo K (written as V/K) may be identified (via the function π) with the range $R(A)$. Such an identification is known as an *isomorphism*. Isomorphic objects are "essentially" the same; they differ only in notation.

5.5 Linear Transformations

Suppose we have a system (the engineer's celebrated "black box") which accepts vectors u from a vector space V as inputs, and produces outputs $v = T(u)$ belonging to a vector space W. Then the system is described by the function $T : V \to W$. A highly desirable property of the system is *linearity*. Physically, this means that the system obeys the *superposition principle*:

(1) The response to a sum of two inputs is the sum of the responses, and
(2) The response to a constant times a given input is that constant times the original response.

Mathematically, (1), which is sometimes called *additivity*, says that $T(u + v) = T(u) + T(v)$, and (2), sometimes called *homogeneity*, says that $T(cu) = cT(u)$. We can combine the two statements into a single requirement, which defines a *linear transformation* from V to W:

If u and v are arbitrary vectors in V and a and b are arbitrary scalars belonging to the underlying field F, then

$$T(au + bv) = aT(u) + bT(v). \tag{3}$$

There is no shortage of examples; in fact, any matrix determines a linear transformation. For if A is an m by n matrix and we take $T(x) = Ax$, $x \in F^n$, then T is a linear transformation from F^n to F^m. Property (1) follows from the distributive law (see property (M4) in Section 5.1), and property (2) follows from the definition of matrix multiplication.

We will see shortly that any linear transformation on a finite-dimensional vector space can be represented by a matrix. But first we will use the linear transformation viewpoint to gain further insights about matrices.

5.5.1 Definitions and Comments. If T is a linear transformation from V to W, the *kernel* of T (notation $\ker T$) is the set of all vectors x in V such that $T(x) = 0$. If T arises from a matrix A, then the kernel of T is the null space $N(A)$, which we discussed in (5.4.4). The *image* of T (notation $\operatorname{im} T$) is the set of all outputs, that is, $\{T(x) : x \in V\}$. (This is consistent with our definition of the image of an arbitrary function in Section 1.5.) If T is associated with the matrix A, then the image of T is the range $R(A)$; again see (5.4.4).

We often write Tx instead of $T(x)$ for elements in the image of T.

We may now add to the six conditions equivalent to invertibility in (5.4.3).

5.5.2 Theorem. *Let A be an n by n matrix, and let T be the associated linear transformation. The following conditions are equivalent to invertibility of A:*

(7) *For any $b \in F^n$, the equations $Ax = b$ have a unique solution (namely, $x = A^{-1}b$).*
(8) *The equations $Ax = 0$ have only the trivial solution $x = 0$.*
(9) *T is injective.*
(10) *The kernel of T is $\{0\}$.*
(11) *T is surjective.*

Proof. If A is invertible and $Ax = b$, multiply on the left by A^{-1} to conclude that x must be $A^{-1}b$, and therefore (7) holds.

(7) implies (8): Take $b = 0$.

(8) implies (9): If $T(u) = T(v)$, then $T(u - v) = 0$, so $u - v$ belongs to $\ker T = N(A)$, and therefore $u = v$.

(9) implies (10): If $T(u) = 0$, then since T is injective and $T(0) = 0$, we have $u = 0$.

(10) implies (11): By hypothesis, the null space of A has dimension 0, so by (5.4.5), the range of A has dimension n. *But an n-dimensional subspace W of an n-dimensional space V must coincide with V.* (For by (5.3.3(d)), any basis for W must also be a basis for V.) Thus the range of A ($=$ the image of T) is all of F^n.

Finally, if T is surjective then the range of A is n-dimensional, so by (5.4.5), A has rank n. By (5.4.3), this is equivalent to invertibility of A. ∎

Now suppose that we have an $n \times p$ matrix B and its associated linear transformation S; then $S(x) = Bx$, $x \in F^p$. Now Bx belongs to F^n, so if A is an $m \times n$ matrix, we have $ABx \in F^m$. What we are actually doing here is composing linear transformations. If T is the linear transformation determined by A, then

$$T\big(S(x)\big) = A(Bx) = ABx,$$

and consequently the product AB corresponds to the composition $T \circ S$. This gives us useful information about the rank of a product.

5.5.3 Theorem.
(a) $\operatorname{rank}(AB) \leq \operatorname{rank} A$.
(b) *If B is invertible then* $\operatorname{rank}(AB) = \operatorname{rank} A$.
(c) $\operatorname{rank}(AB) \leq \operatorname{rank} B$.
(d) *If A is invertible then* $\operatorname{rank}(AB) = \operatorname{rank} B$.

Proof. The range of AB is the image of $T \circ S$, which is contained in the image of T. (Note that $(T \circ S)(x) = T\big(S(x)\big) \in \operatorname{im} T$.) But the image of T is the range of A, so we have $R(AB) \subseteq R(A)$. Since the rank is the dimension of the range (by (5.4.5)), (a) is proved.

If B is invertible, then by (5.5.2), S is surjective, so the image of $T \circ S$ must be the same as the image of T, which proves (b).

To prove (c), we switch to transposes. By (5.4.2), the rank of the transpose of a matrix is the same as the rank of the original matrix. Thus

$$\operatorname{rank}(AB) = \operatorname{rank}\big[(AB)^t\big] = \operatorname{rank}(B^t A^t) \leq \operatorname{rank} B^t \qquad \text{by (a)}$$

$$= \operatorname{rank} B.$$

Finally, if A is invertible, then so is A^t by (5.2.2(b)). Thus in the proof of (c) we have $\operatorname{rank}(B^t A^t) = \operatorname{rank} B^t$ by (b), and consequently $\operatorname{rank}(AB) = \operatorname{rank} B^t = \operatorname{rank} B$. ∎

We are now in a position to discuss a fundamental idea.

5.5.4 The Representation of a Linear Transformation by a Matrix. Suppose that T is a linear transformation from the n-dimensional vector space V to the m-dimensional vector space W. Let $\{v_1, \ldots, v_n\}$ be a basis for V, and $\{w_1, \ldots, w_m\}$ a basis for W. Since $T(v_1)$, which we will now abbreviate to Tv_1, belongs to W, Tv_1 is a linear combination of the w_i, say

$$Tv_1 = a_{11}w_1 + a_{21}w_2 + \cdots + a_{m1}w_m.$$

The column vector C_1 whose components are $a_{11}, a_{21}, \ldots, a_{m1}$ will be the first column of the matrix A that represents T with respect to the two given bases. Similarly,

$$Tv_2 = a_{12}w_1 + a_{22}w_2 + \cdots + a_{m2}w_m,$$

and the column vector with components $a_{21}, a_{22}, \ldots, a_{m2}$ will be the second column of A. In general,

$$Tv_j = a_{1j}w_1 + a_{2j}w_2 + \cdots + a_{mj}w_m = \sum_{i=1}^{m} a_{ij}w_i, \qquad (\alpha)$$

and the column vector with components $a_{1j}, a_{2j}, \ldots, a_{mj}$ will be the jth column of A. (Thus A is an m by n matrix.)

Now suppose that x is a vector in V whose coordinates with respect to the basis $\{v_1, \ldots, v_n\}$ are $c_1, \ldots, c_n$; in other words,

$$x = c_1 v_1 + \cdots + c_n v_n. \qquad (\beta)$$

Then

$$Tx = c_1 Tv_1 + \cdots + c_n Tv_n. \qquad (\gamma)$$

If w is a row vector whose components are w_i, $i = 1, \ldots, m$, Tv is a row vector whose components are Tv_j, $j = 1, \ldots, n$, and c is a column vector whose components are c_i, $i = 1, \ldots, n$, then $Tx = (Tv)c$ by (γ) and $Tv = wA$ by (α); thus

$$Tx = w(Ac) = w_1(Ac)_1 + \cdots w_m(Ac)_m. \qquad (\delta)$$

If we translate Equations (α), (β), (γ), and (δ) into words, we have the following basic result.

> Let $T : V \to W$ be a linear transformation, and let A be the matrix that represents T with respect to the bases $\mathbf{v} = \{v_1, \ldots, v_n\}$ and $\mathbf{w} = \{w_1, \ldots, w_m\}$ for V and W respectively. (In other words, A is the matrix whose jth column contains the coordinates of Tv_j with respect to the w_i.) If c is a column vector which gives the coordinates of x with respect to the v_j, then the coordinates of Tx with respect to the w_i are the components of Ac.

Thus once we choose coordinate systems (that is, bases) for V and W, the matrix A becomes a concrete realization of the abstract linear transformation T. A vector x in V is completely specified by its coordinates c with respect to $\mathbf{v}$, and then Ac gives the coordinates of Tx with respect to $\mathbf{w}$, which completely specifies Tx. The abstract mapping $x \to Tx$ is replaced by the matrix calculation $c \to Ac$. This viewpoint can be exploited when considering the composition of a linear transformation $T_1 : V \to W$ and another linear transformation $T_2 : W \to U$. Suppose that the matrix A_1 represents T_1 with respect to the bases $\mathbf{v}$ and $\mathbf{w}$, and A_2 represents T_2 with respect to $\mathbf{w}$ and $\mathbf{u}$. In order to find the coordinates of $T_2(T_1 x)$, we would first calculate $A_1 c$ and then apply A_2, that is, compute $A_2 A_1 c$. Therefore $T_2 \circ T_1$ is represented by the product $A_2 A_1$. Similarly, the sum of two linear transformations from V to W is represented by the sum of the corresponding matrices.

Let T be a linear transformation from V to W, where V and W have the same dimension (so that the representing matrix A is square). If T is injective, then $Tx = 0$ implies that $x = 0$. This condition translates to $Ac = 0$ implies $c = 0$, which by (5.5.2) part (8) means that A is invertible. Conversely, if A is invertible, then $Tx = 0$ implies $x = 0$, so that T is injective. (If $Tx = Ty$, then $T(x - y) = 0$, so $x = y$.) Similarly, invertibility of A is equivalent to surjectivity of T.

There is a subtle point here that mathematicians like to emphasize. If T is bijective, then there is an inverse function T^{-1} mapping backwards from W to V. There may be a tendency to assume automatically that T^{-1} is a linear transformation, represented by the matrix A^{-1}, but this must be proved (see Problem 1).

The correspondence between linear transformations and matrices allows us to restate the Dimension Theorem (5.4.5); see Problem 7 for an application.

5.5.5 Dimension Theorem For Linear Transformations. *Let T be a linear transformation from an n-dimensional vector space V to an m-dimensional vector space W. Then $\dim(\ker T) + \dim(\operatorname{im} T) = n$. In fact, if $u_1, \ldots, u_k$ form a basis for $\ker T$, and the u_i are extended to a basis $u_1, \ldots u_n$ for V, then $Tu_{k+1}, \ldots, Tu_n$ form a basis for $\operatorname{im} T$.* ∎

We now investigate what happens to matrices and coordinates when the basis is changed. First suppose that we replace an old basis $\mathbf{v} = \{v_1, \ldots, v_n\}$ of V by a new basis $\mathbf{v}' = \{v_1', \ldots, v_n'\}$. Let P be the *basis-changing matrix,* that is, the matrix whose jth column contains the coordinates of v_j' with respect to the v_i. The crucial observation is that P is the matrix that represents the identity transformation $I : V \to W = V$ with respect to the bases $\mathbf{v}'$ and $\mathbf{v}$; to see this, note that the jth column of the representing matrix gives the coordinates of $Iv_j' \ (= v_j')$ with respect to the v_i. Thus the descriptions of the representing matrix and of the matrix P are identical.

But by (5.5.4), P times the column vector of coordinates of x with respect to $\mathbf{v}'$ gives the coordinates of $x \ (= Ix)$ with respect to $\mathbf{v}$. We may state this result as follows.

5.5.6 Coordinates Under Change of Basis. If P is the basis-changing matrix corresponding to a switch from an old basis $\mathbf{v}$ to a new basis $\mathbf{v}'$, then

$$P \text{ (new coordinates)} = \text{old coordinates.}$$

Therefore

$$P^{-1} \text{ (old coordinates)} = \text{new coordinates.}$$

(Note that P is invertible because the identity transformation is injective.)

Now let T be a linear transformation from V to W, and suppose we change bases from $\mathbf{v}$ to $\mathbf{v}'$ in V, and from $\mathbf{w}$ to $\mathbf{w}'$ in W. Let P be the basis-changing matrix on V, and Q the basis-changing matrix on W. If A is the matrix that represents T with respect to $\mathbf{v}$ and $\mathbf{w}$, and B is the matrix representing T with respect to $\mathbf{v}'$ and $\mathbf{w}'$, we have the following relation between A and B.

5.5.7 Matrices Under Change of Basis. *If P and Q are the basis-changing matrices on V and W respectively, A represents T with respect to the old bases, and B represents T with respect to the new bases, then*

$$B = Q^{-1}AP.$$

Proof. Let c and c' be the column vectors of coordinates of x with respect to the old and new bases, respectively. By (5.5.6), $c = Pc'$. The coordinates of Tx with respect to the old basis are given by $Ac = APc'$. The new coordinates of Tx are given by Q^{-1} times the old coordinates; that is, $Q^{-1}Ac = (Q^{-1}AP)c'$. Thus the matrix that operates on the coordinates of x and produces the coordinates of Tx (with respect to the new bases) is $B = Q^{-1}AP.$ ∎

Most of the time we will be assuming that $V = W$, $\mathbf{v} = \mathbf{w}$, and $\mathbf{v}' = \mathbf{w}'$, so that we are switching from an old basis $\mathbf{v}$ on V to a new basis $\mathbf{v}'$. In this case we say that T is a *linear operator* on V and that A represents T with respect to the basis $\mathbf{v}$. Then A is a square matrix, $Q = P$, and

$$B = P^{-1}AP.$$

Square matrices A and B such that $B = P^{-1}AP$ for some nonsingular P are said to be *similar.* They represent the same linear operator on V with respect to different bases.

5.5.8 Example. Let T be the linear operator on $\mathbb{R}^3$ determined by

$$Te_1 = (2,0,0), \qquad Te_2 = (0,0,1), \qquad Te_3 = (0,1,0)$$

where $\{e_1, e_2, e_3\}$ is the standard basis. The matrix that represents T with respect to the standard basis is

$$A = \begin{bmatrix} 2 & 0 & 0 \\ 0 & 0 & 1 \\ 0 & 1 & 0 \end{bmatrix}.$$

Suppose we change to a new basis $u = (1,0,0)$, $v = (1,1,0)$, $w = (0,0,2)$. The basis-changing matrix P has columns that give the coordinates of the new basis vectors with respect to the old basis. Thus

$$P = \begin{bmatrix} 1 & 1 & 0 \\ 0 & 1 & 0 \\ 0 & 0 & 2 \end{bmatrix} \qquad \text{and} \qquad P^{-1} = \begin{bmatrix} 1 & -1 & 0 \\ 0 & 1 & 0 \\ 0 & 0 & 1/2 \end{bmatrix}.$$

The matrix of T with respect to the new basis is

$$B = P^{-1}AP = \begin{bmatrix} 2 & 2 & -2 \\ 0 & 0 & 2 \\ 0 & 1/2 & 0 \end{bmatrix}.$$

Since the coordinates of the vector (a, b, c) with respect to the standard basis are simply a, b, and c, the matrix A gives us a formula for T:

$$T(a, b, c) = (2a, c, b).$$

Now if $x = au + bv + cw$, the matrix B is used to find the coordinates of Tx with respect to the new basis:

$$T(au + bv + cw) = (2a + 2b - 2c)u + 2cv + \tfrac{1}{2}bw.$$

Problems For Section 5.5

1. If T is a bijective linear transformation represented by the matrix A, show that T^{-1} is also a linear transformation, and is represented by A^{-1}.

2. Suppose we change bases for $\mathbb{R}^2$ from the standard basis $\{e_1, e_2\}$ to $\{u, v\}$, and the old coordinates x, y are related to the new coordinates X, Y by

$$X = 4x - 6y$$
$$Y = y.$$

Find the basis-changing matrix P, and express u and v in terms of e_1 and e_2.

3. If $T : \mathbb{R}^2 \to \mathbb{R}^2$ is counterclockwise rotation by the angle θ, find the matrix of T with respect to the standard basis.

4. Let L be a line through the origin in the plane, and assume that the slope of L is a. Then the vector $u = (1, a)$, when pictured as an arrow from the origin to $(1, a)$, is parallel to L, and the vector $v = (-a, 1)$ is perpendicular to L. Let T be the linear transformation that reflects points about the line L.

Find the matrix B of T with respect to the basis $\{u, v\}$, and from this find the matrix A of T with respect to the standard basis.

5. Show that similar matrices have the same rank.

6. Show that if A is similar to B, then A^t is similar to B^t.

7. Let $T : V \to W$ be a linear transformation.
    (a) If $\dim V > \dim W$, show that T cannot be injective.
    (b) If $\dim V < \dim W$, show that T cannot be surjective.

5.6 Inner Product Spaces

If $x = (a_1, a_2, a_3)$ and $y = (b_1, b_2, b_3)$ are vectors in $\mathbb{R}^3$, the *dot product* of x and y is familiar from calculus:

$$x \cdot y = a_1 b_1 + a_2 b_2 + a_3 b_3.$$

The dot product gives us many geometric ideas, for example, the length of a vector:

$$||x|| = (x \cdot x)^{1/2} = \left[(a_1)^2 + (a_2)^2 + (a_3)^2 \right]^{1/2}$$

and the angle θ between two vectors:

$$x \cdot y = ||x|| \, ||y|| \cos \theta.$$

The component of a vector x in the direction of a vector u is

$$x \cdot \left(\frac{u}{||u||} \right)$$

which is the dot product of x and a unit vector (a vector of length 1) in the direction of u.

We can generalize the dot product to n dimensions, and it will be very useful to allow complex as well as real components. So from now on, we take the underlying field F to be the complex numbers $\mathbb{C}$. A vector space over $\mathbb{C}$ is sometimes called a *complex vector space* (and a vector space over $\mathbb{R}$ is called a *real vector space*).

5.6.1 Definitions and Comments. Let V be a vector space over the complex numbers. An *inner product* on V is a function that assigns to each pair (x, y) of vectors in V a complex number $\langle x, y \rangle$ satisfying, for all $x, y, z \in V$ and all $a, b \in \mathbb{C}$,

$$\langle ax + by, z \rangle = a \langle x, z \rangle + b \langle y, z \rangle \tag{1}$$

(the inner product is linear in the first variable);

$$\langle x, x \rangle \text{ is real and } \geq 0, \text{ with equality if and only if } x = 0; \tag{2}$$

$$\langle y, x \rangle = \overline{\langle x, y \rangle}, \text{ the complex conjugate of } \langle x, y \rangle. \tag{3}$$

It follows from (1) and (3) that

$$\langle x, ay + bz \rangle = \overline{a} \langle x, y \rangle + \overline{b} \langle x, z \rangle \tag{4}$$

(the inner product is conjugate linear in the second variable).

The *norm* or *length* of a vector $x \in V$ is defined as

$$||x|| = \sqrt{\langle x, x \rangle}.$$

By (1) and (4) we have $||ax||^2 = \langle ax, ax \rangle = a\overline{a}||x||^2 = |a|^2 ||x||^2$, so

$$||ax|| = |a| \, ||x|| \tag{5}$$

By (2) we have

$$||x|| \geq 0, \text{ with equality if and only if } x = 0. \tag{6}$$

Two vectors x and y are said to be *orthogonal* or *perpendicular* (notation $x \perp y$) if $\langle x, y \rangle = 0$. When we say that more than two vectors, say $x_1, \ldots, x_k$, are orthogonal, we always mean that they are *mutually perpendicular*; in other words, $x_i \perp x_j$ whenever $i \neq j$. The vectors $x_1, \ldots, x_k$ are said to be *orthonormal* (or the set $\{x_1, \ldots, x_k\}$ is orthonormal) if the vectors are orthogonal *unit vectors* (that is, $||x_i|| = 1$ for all i).

Nonzero orthogonal vectors $x_1, \ldots, x_k$ must be linearly independent. For if $a_1 x_1 + \cdots + a_k x_k = 0$, take the inner product of both sides with x_i to get $a_i \langle x_i, x_i \rangle = 0$, and since $x_i \neq 0$, we must have $a_i = 0$.

An *inner product space* is a vector space V with an inner product $\langle \ \rangle$ defined on V. Our canonical example will be $\mathbb{C}^n$, the set of all n-tuples with components in $\mathbb{C}$, with

$$\langle (a_1, \ldots, a_n), (b_1, \ldots, b_n) \rangle = a_1 \overline{b_1} + \cdots + a_n \overline{b_n}.$$

In fact this is essentially the only n-dimensional example we need to consider; we will clarify this idea after we have studied orthonormal bases.

We now derive some basic properties of inner products and norms. In $\mathbb{R}^2$ or $\mathbb{R}^3$, these results have a familiar geometric interpretation. The Cauchy-Schwarz Inequality says that the cosine of the angle between two vectors is between -1 and $+1$. The Triangle Inequality states that the length of one side of a triangle is less than or equal to the sum of the lengths of the other two sides. The Parallelogram Law says that the sum of the squares of the lengths of the diagonals of a parallelogram is twice the sum of the squares of the lengths of the sides. The Pythagorean Theorem asserts (when $k = 2$ in (5.6.5) below) that the square of the length of the hypotenuse of a right triangle equals the sum of the squares of the lengths of the sides.

5.6.2 Cauchy-Schwarz Inequality. $|\langle x, y \rangle| \leq ||x|| \, ||y||.$

Proof.   For any complex number a,

$$0 \leq \langle x + ay, x + ay \rangle = \langle x + ay, x \rangle + \langle x + ay, ay \rangle$$

$$= \langle x, x \rangle + a \langle y, x \rangle + \overline{a} \langle x, y \rangle + |a|^2 \langle y, y \rangle.$$

Set $a = -\langle x, y \rangle / \langle y, y \rangle$ (if $\langle y, y \rangle = 0$ then $y = 0$ and both sides of the Cauchy-Schwarz inequality are 0). Since $\langle y, x \rangle$ is the conjugate of $\langle x, y \rangle$, we have

$$0 \leq \langle x, x \rangle - 2 \frac{|\langle x, y \rangle|^2}{\langle y, y \rangle} + \frac{|\langle x, y \rangle|^2}{\langle y, y \rangle}.$$

Since $\langle x, x \rangle = ||x||^2$ and $\langle y, y \rangle = ||y||^2$, the result follows. ∎

5.6.3 Triangle Inequality. $||x + y|| \leq ||x|| + ||y||.$

Proof.   As in the proof of (5.6.2),

$$||x + y||^2 = \langle x + y, x + y \rangle = \langle x, x \rangle + \langle x, y \rangle + \langle y, x \rangle + \langle y, y \rangle$$

$$= ||x||^2 + ||y||^2 + 2\text{Re}\langle x, y \rangle$$

since the sum of a complex number c and its conjugate $\bar{c}$ is twice the real part of c

$$\leq ||x||^2 + ||y||^2 + 2|\langle x, y \rangle|$$

since the real part of a complex number c is at most equal to the magnitude of c

$$\leq ||x||^2 + ||y||^2 + 2||x||\,||y|| \qquad \text{by (5.6.2)}.$$

Take the square root of both sides to complete the proof. ∎

5.6.4 Parallelogram Law. $||x + y||^2 + ||x - y||^2 = 2(||x||^2 + ||y||^2)$.

Proof.   As in the proof of (5.6.3), we have

$$||x + y||^2 = ||x||^2 + ||y||^2 + 2\text{Re}\langle x, y \rangle \qquad \text{and}$$
$$||x - y||^2 = ||x||^2 + ||y||^2 - 2\text{Re}\langle x, y \rangle;$$

add these equations to finish the argument. ∎

5.6.5 Pythagorean Theorem.   If $x_1, x_2, \ldots, x_k$ are orthogonal, then

$$||x_1 + \cdots + x_k||^2 = ||x_1||^2 + \cdots + ||x_k||^2.$$

Proof.   As in the proof of (5.6.3),

$$\left\| \sum_{i=1}^{k} x_i \right\|^2 = \sum_{i=1}^{k} ||x_i||^2 + 2 \sum_{i<j} \text{Re}\langle x_i, x_j \rangle,$$

and since $\langle x_i, x_j \rangle = 0$ for $i \neq j$, the result follows. ∎

Now suppose that $\{x_1, \ldots, x_k\}$ is an *orthonormal basis* for a subspace S of V, in other words, a basis for S consisting of orthonormal vectors. If x is a vector in S, the coordinates of x with respect to the x_i are the inner products $\langle x, x_i \rangle$.

5.6.6 Theorem.   *If x belongs to the subspace S and $\{x_1, \ldots, x_k\}$ is an orthonormal basis for S, then $x = \sum_{i=1}^{k} \langle x, x_i \rangle x_i$. Consequently, by (5.6.5),*

$$||x||^2 = \sum_{i=1}^{k} |\langle x, x_i \rangle|^2.$$

Proof.   We have $x = a_1 x_1 + \cdots + a_k x_k$ for some scalars $a_1, \ldots, a_k$. Take the inner product of both sides with x_i to obtain, using the orthonormality of $\{x_1, \ldots, x_k\}$, $\langle x, x_i \rangle = a_i$. ∎

Now suppose that x is any vector in V, not necessarily in S, and we wish to find the *projection* of x on S, that is, the vector in S that is closest to x. Thus among all vectors $a_1 x_1 + \cdots + a_k x_k$ in S, we must choose the a_i so that $x - \sum_{i=1}^{k} a_i x_i$ has minimum length. The result is very explicit.

5.6.7 Theorem. *If $\{x_1, \ldots, x_k\}$ is an orthonormal basis for a subspace S and x is an arbitrary vector, then $\|x - \sum_{i=1}^{k} a_i x_i\|$ is minimized when $a_i = \langle x, x_i \rangle$, $i = 1, \ldots, k$. The coefficients a_i that achieve the minimum are unique.*

Proof. By the now familiar calculation of (5.6.3),

$$\left\| x - \sum_{i=1}^{k} a_i x_i \right\|^2 = \left\langle x - \sum_{i=1}^{k} a_i x_i, x - \sum_{j=1}^{k} a_j x_j \right\rangle$$

$$= \|x\|^2 - \sum_{j=1}^{k} \overline{a}_j \langle x, x_j \rangle - \sum_{i=1}^{k} a_i \langle x_i, x \rangle + \left\langle \sum_{i=1}^{k} a_i x_i, \sum_{j=1}^{k} a_j x_j \right\rangle$$

and the last term on the right is $\sum_{i=1}^{k} |a_i|^2$ since the x_i are orthonormal. Furthermore,

$$-\overline{a}_i \langle x, x_i \rangle - a_i \langle x_i, x \rangle + |a_i|^2 = -|\langle x, x_i \rangle|^2 + |a_i - \langle x, x_i \rangle|^2,$$

as can be seen by writing the last term on the right as

$$\bigl(a_i - \langle x, x_i \rangle\bigr)\bigl(\overline{a}_i - \langle x_i, x \rangle\bigr).$$

Thus

$$0 \le \left\| x - \sum_{i=1}^{k} a_i x_i \right\|^2 = \|x\|^2 - \sum_{i=1}^{k} |\langle x, x_i \rangle|^2 + \sum_{i=1}^{k} |a_i - \langle x, x_i \rangle|^2,$$

so to achieve a minimum we have no choice but to take $a_i = \langle x, x_i \rangle$ for all $i = 1, \ldots, k$. ∎

If S is any subspace of a finite-dimensional vector space V, we know that S has a basis (see Section 5.3), and in fact if V is an inner product space, we can always produce an orthonormal basis for S.

5.6.8 Gram-Schmidt Process. Let $x_1, \ldots, x_r$ be arbitrary linearly independent vectors. We are going to find (inductively) orthonormal vectors $y_1, \ldots, y_r$ that span exactly the same space as the x_i. We start with $y_1 = x_1 / \|x_1\|$, and certainly y_1 spans the same space as x_1. Suppose we have chosen orthonormal $y_1, \ldots, y_n$ so that $y_1, \ldots, y_n$ span the same subspace S as $x_1, \ldots, x_n$. The idea is to replace x_{n+1} by x_{n+1} minus the projection p_{n+1} of x_{n+1} on S (see Fig. 5.6.1).

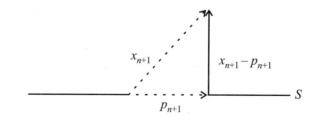

FIGURE 5.6.1
Gram-Schmidt Process

We normalize by taking

$$y_{n+1} = \frac{x_{n+1} - p_{n+1}}{||x_{n+1} - p_{n+1}||}.$$

(Note that x_{n+1} cannot belong to S, by linear independence of the x_i; thus there is no division by 0.) Now by (5.6.7),

$$p_{n+1} = \sum_{i=1}^{n} \langle x_{n+1}, y_i \rangle y_i,$$

and if we take the inner product of both sides with y_i, we find that $\langle p_{n+1}, y_i \rangle = \langle x_{n+1}, y_i \rangle$; in other words, $x_{n+1} - p_{n+1}$ is orthogonal to y_i, $i = 1, \ldots, n$. Thus $y_1, \ldots, y_{n+1}$ are orthonormal. Now since $p_{n+1} \in S$, the subspace spanned by $x_1, \ldots, x_n$, it follows that y_{n+1} is a linear combination of the x_i, $i = 1, \ldots, n+1$. Finally, since x_{n+1} is a linear combination of p_{n+1} and y_{n+1}, x_{n+1} is a linear combination of the y_i, $i = 1, \ldots, n+1$. Thus $x_1, \ldots, x_{n+1}$ and $y_1, \ldots, y_{n+1}$ span the same space, completing the induction.

Let us extract the key idea in the above computation. *If S is a subspace and p is the projection of a vector x on S, then $x - p$ is orthogonal to S,* that is, orthogonal to every vector in S. (Notation $x - p \perp S$.) For we know that S has an orthonormal basis, say $y_1, \ldots, y_n$. The above argument, with x_{n+1} replaced by x and p_{n+1} by p, shows that $x - p$ is orthogonal to each y_i. But every vector in S is a linear combination of the y_i, and by linearity of the inner product, $x - p$ is orthogonal to everything in S. (This conclusion holds even if x is in S to begin with, since in this case $p = x$.)

If a vector y belongs to S and $x - y$ is orthogonal to S, then y must be the projection of x on S, as we now prove.

5.6.9 Projection Theorem. *Let S be a subspace of the finite-dimensional inner product space V, and let x be an arbitrary vector in V. Then x can be expressed uniquely as $x = y + z$, where $y \in S$ and $z \perp S$. Furthermore, y is the projection of x on S.*

Proof. If p is the projection of x on S, we have just seen that $p \in S$ and $x - p \perp S$; thus we may take $y = p$ and $z = x - p$. Now if we also have $x = y' + z'$ with $y' \in S$ and $z' \perp S$, then $y - y' = z' - z$, and this vector, call it w, is both in S and orthogonal to S. But then w is orthogonal to itself, which implies that $w = 0$. Thus $y = y'$ and $z = z'$. ∎

Geometrically, we can think of x as the hypotenuse of a right triangle with sides y and z, and because $z \perp S$, y is often called the *orthogonal projection* of x on S. We will not need the extra term; for us, projection will always mean orthogonal projection.

We can define a linear operator P on V, called a **projection operator,** by $Px =$ the projection of x on the subspace S. (If $x_1 = y_1 + z_1$, $x_2 = y_2 + z_2$, and $x = ax_1 + bx_2$ where the y_j belong to S and the z_j are orthogonal to S, then $x = (ay_1 + by_2) + (az_1 + bz_2) = y + z$ with $y \in S$, $z \perp S$. Thus $Px = aPx_1 + bPx_2$, proving linearity.) We will study this class of operators in detail in Sections 6.3 and 6.4.

We remarked at the beginning of the section that $\mathbb{C}^n$ is the canonical inner product space. To justify this assertion, suppose that V is any n-dimensional inner product space. The Gram-Schmidt process guarantees the existence of an orthonormal basis $x_1, \ldots, x_n$. If

x is any vector in V, then $x = \sum_{i=1}^{n} a_i x_i$ where $a_i = \langle x, x_i \rangle$ by (5.6.6). We can identify x with the n-tuple $(a_1, \ldots, a_n)$ of coordinates, and in fact the mapping $x \to (a_1, \ldots, a_n)$ is a bijective linear transformation from V to $\mathbb{C}^n$. Furthermore, inner products are preserved: if

$$x = \sum_{i=1}^{n} a_i x_i \qquad \text{and} \qquad y = \sum_{i=1}^{n} b_i x_i,$$

then by orthonormality of the x_i we have

$$\langle x, y \rangle = \sum_{i=1}^{n} a_i \bar{b}_i.$$

Thus V and $\mathbb{C}^n$ are essentially two copies of the same space; they differ only notationally. We say that the mapping $x \to (a_1, \ldots, a_n)$ is an *isomorphism* between V and $\mathbb{C}^n$.

5.6.10 Example. Let $u = (1, 0, 0, 0)$, $v = (1, 0, 1, 0)$, $w = (0, 1, 0, 2)$, $x = (3, 2, 1, 2)$. Find the projection p of x on the subspace S of $\mathbb{C}^4$ spanned by u, v, and w.

Write p as a linear combination of u, v, and w:

$$p = au + bv + cw = a(1, 0, 0, 0) + b(1, 0, 1, 0) + c(0, 1, 0, 2) = (a + b, c, b, 2c).$$

Then $x - p = (3 - a - b, 2 - c, 1 - b, 2 - 2c)$. By (5.6.9), $x - p \perp S$; that is, $x - p \perp u$, v, and w. This gives us three equations for a, b, and c:

$$3 - a - b = 0, \qquad 3 - a - b + 1 - b = 0, \qquad 2 - c + 4 - 4c = 0.$$

Simplifying, we have

$$\begin{aligned}
a + b &= 3 \\
a + 2b &= 4 \\
5c &= 6.
\end{aligned}$$

The solution is $a = 2$, $b = 1$, $c = \frac{6}{5}$, so that $p = 2u + v + \frac{6}{5}w = \left(3, \frac{6}{5}, 1, \frac{12}{5}\right)$.

A slick approach to this problem is to form the matrix A whose columns are u, v, and w. If q is the column vector with components a, b, c, then q may be found by solving the equations $A^t A q = A^t x$, which indeed are the above three equations for a, b, and c. See Problem 4 for details.

If we were to ask for the projection of x on the subspace of $\mathbb{R}^4$ (rather than $\mathbb{C}^4$) spanned by u, v, and w, the result would be the same, since only real numbers are involved in the computation. In fact all the theory developed in this section is valid for real vector spaces; simply delete all overbars (indicating complex conjugation). But beginning with the next section, complex scalars will be essential (see (5.7.2)).

Problems For Section 5.6

1. Prove the *Polarization Identity*: in any inner product space,

$$4\langle x, y \rangle = ||x + y||^2 - ||x - y||^2 + i||x + iy||^2 - i||x - iy||^2 \qquad \text{where } i = \sqrt{-1}.$$

Thus if we know the norm of each vector in an inner product space, the inner product is determined.

2. If $x_1, \ldots, x_k$ are orthonormal and x is any vector, prove *Bessel's Inequality*:

$$||x||^2 \geq \sum_{i=1}^{k} |\langle x, x_i \rangle|^2.$$

3. If S is any subset of the inner product space V, let $S^\perp$ be the set of vectors orthogonal to S. Show that S is a subspace of V.

4. Let $x_1, \ldots, x_k$ span a subspace S of $\mathbb{R}^n$, and let $p = a_1 x_1 + \cdots + a_k x_k$ be the projection of a (column) vector x on S. If A is the matrix whose columns are $x_1, \ldots, x_k$ and q is a column vector whose components are $a_1, \ldots, a_k$, show that p may be found by solving the *normal equations*

$$A^t A q = A^t x.$$

What is the form of the normal equations if $\mathbb{R}^n$ is replaced by $\mathbb{C}^n$?

5. (*Least Squares*) Suppose that y_i is a (real) numerical measurement taken at time x_i, $i = 1, \ldots, m$, and we wish to fit the data to a line $y = ax + b$, so that the sum of squares of vertical distances from the data points to the line, namely,

$$E = \sum_{i=1}^{m} |y_i - ax_i - b|^2,$$

is minimized. Show that this problem is equivalent to projecting the vector Y whose components are the y_i on the subspace spanned by $(x_1, \ldots, x_m)$ and $(1, \ldots, 1)$. If

$$X = \begin{bmatrix} a \\ b \end{bmatrix},$$

show that the normal equations for this problem are

$$A^t A X = A^t Y$$

where A is an $m \times 2$ matrix whose columns are $(x_1, \ldots, x_m)$ and $(1, \ldots, 1)$.

6. Continuing Problem 5, indicate how to solve the least squares problem if a parabola $y = ax^2 + bx + c$ is fitted to the data.

7. Under what conditions will there be equality in the Cauchy-Schwarz Inequality 5.6.2?

5.7 Eigenvalues and Eigenvectors

As in the previous section, we take the underlying field to be $\mathbb{C}$. We are now going to study linear transformations $T : V \to W$ where V and W coincide; such transformations are called *linear operators* on V. In return for this restriction, we gain a key idea, that of a *polynomial* in the linear operator T.

5.7.1 Definitions and Comments. If T is a linear operator on the n-dimensional vector space V, then by T^2 we mean $T \circ T$, the composition of T with itself. Similarly, T^3 is

$T \circ T \circ T$, and in general, T^k is $T \circ T \circ \cdots \circ T$ (with k occurrences of T). If $p(x) = a_m x^m + \cdots + a_1 x + a_0$ is any polynomial, we define $p(T) = a_m T^m + \cdots + a_1 T + a_0 I$, where I is the identity operator ($I(y) = y$ for all $y \in V$).

If T is represented by the matrix A with respect to a given basis, then A is n by n, and by the discussion in (5.5.4), $p(T)$ is represented by $p(A) = a_m A^m + \cdots + a_1 A + a_0 I$.

We are going to look for a basis $x_1, \ldots, x_n$ in which computations involving T can be done with minimal effort. A promising candidate is a basis such that T simply multiplies each x_i by a constant λ_i; thus $Tx_i = \lambda_i x_i$, $i = 1, \ldots, n$. It follows from (5.5.4) that the matrix A of T with respect to this basis has zeros off the main diagonal (in other words, $a_{ij} = 0$ for all $i \neq j$). Furthermore, $a_{ii} = \lambda_i$, $i = 1, \ldots, n$. Such an A is called a *diagonal matrix,* and we write $A = \mathrm{diag}(\lambda_1, \ldots, \lambda_n)$. Thus if the coordinates of x are given by $c = (c_1, \ldots, c_n)$, then the coordinates of Tx are given by $Ac = (\lambda_1 c_1, \ldots, \lambda_n c_n)$. (In general, if a matrix B is premultiplied, that is, multiplied on the left, by $\mathrm{diag}(\lambda_1, \ldots, \lambda_n)$, then for each i, row i of B is multiplied by λ_i. If B is postmultiplied (multiplied on the right) by $\mathrm{diag}(\lambda_1, \ldots, \lambda_n)$, then for each i, column i of B is multiplied by λ_i.)

We are led to consider the following problem.

5.7.2 The Eigenvalue Problem.

Let T be a linear operator on the finite-dimensional vector space V. If $Tx = \lambda x$ for some $\lambda \in \mathbb{C}$ and some nonzero $x \in V$, we say that λ is an *eigenvalue* of T and that x is an *eigenvector* of T. Similarly, if $Ac = \lambda c$, $c \neq 0$, where A is an n by n matrix and c is a column vector with n components, we say that λ is an eigenvalue of A and that c is an eigenvector of A.

We are trying to form a basis consisting entirely of eigenvectors. This turns out to be possible very often, but not always. The equation $Tx = \lambda x$ is equivalent to $(T - \lambda I)x = 0$, or in terms of matrices, $(A - \lambda I)c = 0$. By conditions (2) and (8) of (5.4.3) and (5.5.2), a nontrivial solution will exist if and only if $\det(A - \lambda I) = 0$. Notice that the determinant does not depend on the particular matrix we have chosen to represent T, and consequently *similar matrices have the same eigenvalues.* For if $B = P^{-1}AP$ (see (5.5.7)), then

$$\det(B - \lambda I) = \det(P^{-1}AP - \lambda I) = \det\left(P^{-1}(A - \lambda I)P\right)$$

$$= \det P^{-1} \det(A - \lambda I) \det P = \det(A - \lambda I)$$

since $\det P^{-1} = 1/\det P$. Thus without ambiguity we may refer to the determinant of $T - \lambda I$. Similarly, it follows that if $B = P^{-1}AP$ then $\det B = \det A$, so we are free to define the *determinant of the linear operator T.*

If there is a basis of eigenvectors, we say that the linear operator T (or the representing matrix A) is *diagonalizable,* or that T (or A) *can be diagonalized.*

For example, if

$$A = \begin{bmatrix} 1 & 2 \\ 3 & 4 \end{bmatrix}, \quad \text{then} \quad A - \lambda I = \begin{bmatrix} 1 - \lambda & 2 \\ 3 & 4 - \lambda \end{bmatrix},$$

and the determinant of $A - \lambda I$ is $\lambda^2 - 5\lambda - 2$. In general, $\det(A - \lambda I)$ is a polynomial of degree n, and *since our scalars are complex rather than real,* we may conclude that

a linear operator on an n-dimensional space has n eigenvalues, counting multiplicity.

If there are no repeated roots, that is, if T has n distinct eigenvalues $\lambda_1, \ldots, \lambda_n$, we are going to show that T is diagonalizable. We need one preliminary result.

5.7.3 Theorem. *If $x_1, \ldots, x_k$ are eigenvectors corresponding to distinct eigenvalues $\lambda_1, \ldots, \lambda_k$, then the x_i are linearly independent.*

Proof. If $a_1 x_1 + \cdots + a_k x_k = 0$, apply $p(T) = (T - \lambda_2 I) \cdots (T - \lambda_k I)$ to both sides. Since $(T - \lambda_i I)(T - \lambda_j I) = (T - \lambda_j I)(T - \lambda_i I)$ for all i, j, we have $p(T)x_j = 0$, $j = 2, \ldots, k$. Now

$$(T - \lambda_j I)x_1 = Tx_1 - \lambda_j x_1 = \lambda_1 x_1 - \lambda_j x_1 = (\lambda_1 - \lambda_j)x_1, \qquad j = 2, \ldots, k.$$

Thus

$$p(T)(a_1 x_1 + \cdots + a_k x_k) = a_1(\lambda_1 - \lambda_2)(\lambda_1 - \lambda_3) \cdots (\lambda_1 - \lambda_k)x_1 = 0,$$

and since the λ_j are distinct and x_1 (an eigenvector) is not zero, we must have $a_1 = 0$. We may repeat this argument to show that all the a_i are 0. ∎

5.7.4 Theorem. *A linear operator T (or matrix A) with distinct eigenvalues $\lambda_1, \ldots, \lambda_n$ can be diagonalized. In fact if $D = \operatorname{diag}(\lambda_1, \ldots, \lambda_n)$ and P is the matrix whose columns are $x_1, \ldots, x_n$, where x_i is an eigenvector for the eigenvalue λ_i, then $P^{-1}AP = D$.*

Proof. It's easier to work with A rather than T. Let x_i be an eigenvector of A for the eigenvalue λ_i, $i = 1, \ldots, n$, and let P be the matrix whose columns are the x_i. Thus $P = [x_1 \mid x_2 \; \cdots \; \mid x_n]$, and column i of AP is $Ax_i = \lambda_i x_i$, $i = 1, \ldots, n$. By (5.7.3), $\{x_1, \ldots, x_n\}$ is a basis, so P has n linearly independent columns and is therefore nonsingular. Now if P is postmultiplied by $D = \operatorname{diag}(\lambda_1, \ldots, \lambda_n)$, column i of the resulting matrix PD is λ_i times column i of P, that is, $\lambda_i x_i$. Consequently, $AP = PD$, so that $P^{-1}AP = D$. ∎

The diagonalization of the linear operator T can also be viewed as a decomposition of the space V into subspaces, as follows.

5.7.5 Definitions and Comments. Let V_i be the *eigenspace* of λ_i, that is, the subspace of V consisting of the eigenvectors of λ_i and the zero vector. (V_i is indeed a subspace; if $x, y \in V_i$ and $a, b \in \mathbb{C}$, then $T(ax + by) = aTx + bTy = a\lambda_i x + b\lambda_i y = \lambda_i(ax + by)$, so that $ax + by \in V_i$). Then V_i is one-dimensional and is spanned by x_i. (We are still assuming that T has distinct eigenvalues. If any V_i had dimension greater than 1, we would have more than n linearly independent eigenvectors, which is impossible.) In fact V is the *direct sum* of the V_i, which means that each $y \in V$ can be expressed uniquely as $y_1 + \cdots + y_n$ where $y_i \in V_i$, $i = 1, \ldots, n$. For if $y \in V$, then, since the eigenvectors x_i form a basis, and V_i consists of all multiples of x_i, we have, for some scalars $a_1, \ldots, a_n$, $y = a_1 x_1 + \cdots + a_n x_n = y_1 + \cdots + y_n$ with $y_i \in V_i$, $i = 1, \ldots, n$. If $y_1 + \cdots + y_n = y_1' + \cdots + y_n'$ with $y_i' \in V_i$, then $\sum_{i=1}^{n}(y_i - y_i') = 0$ with $y_i - y_i' \in V_i$, $i = 1, \ldots, n$. The linear independence of the eigenvectors (see (5.7.3)) implies that $y_i = y_i'$ for all i, proving uniqueness. On the subspace V_i, T simply multiplies vectors by λ_i, so

V_i is *T-invariant*; that is, $x \in V_i$ implies $Tx \in V_i$. Furthermore, the restriction of T to V_i is represented by the 1 by 1 matrix $[\lambda_i]$. These 1 by 1 matrices are assembled to form the main diagonal of the diagonal matrix $D = \text{diag}(\lambda_1, \ldots, \lambda_n)$, and we say that D is the direct sum of the matrices $[\lambda_i]$. Direct sums can be somewhat more complicated. For example, if A is a p by p matrix, B is q by q, and C is r by r, we can manufacture a large matrix

$$M = \begin{bmatrix} A & 0 & 0 \\ 0 & B & 0 \\ 0 & 0 & C \end{bmatrix}.$$

Thus A occupies the first p rows and columns of M, B the next q rows and columns, and C the last r rows and columns; the other entries are 0. Let V_1 be the subspace of vectors of the form $(x_A, 0, 0)$ (all entries 0 except possibly in the first p components). Similarly, let V_2 be the subspace of vectors that look like $(0, x_B, 0)$, and V_3 the subspace of vectors of the form $(0, 0, x_C)$. Then V is the direct sum of V_1, V_2, and V_3, and we also say that M is the direct sum of A, B, and C. We use the notation

$$V = V_1 \oplus V_2 \oplus V_3 \ (\text{or } V = \bigoplus_{i=1}^{3} V_i) \quad \text{and} \quad M = A \oplus B \oplus C$$

to indicate direct sums.

A basis of eigenvectors simplifies computations considerably, but if we are working in an inner product space, an *orthonormal* basis of eigenvectors would be even more desirable. It is tempting to argue as follows. Suppose that $x_1, \ldots, x_n$ comprise a basis of eigenvectors, with x_i corresponding to the eigenvalue λ_i. Simply orthonormalize the x_i by the Gram-Schmidt process, and we're finished. The difficulty is that although the orthonormal vectors $y_1, \ldots, y_n$ obtained by Gram-Schmidt span the same space as the x_i, namely all of V, there is no guarantee that the y_i will be eigenvectors. Each y_i is a linear combination of the x_i, but how do we know that $a_1 x_1 + \cdots + a_n x_n$ is an eigenvector? As long as we stay *within the same eigenspace,* we are OK. If $Au = \lambda u$ and $Av = \lambda v$, then $A(au + bv) = \lambda(au + bv)$. If we orthonormalize each eigenspace and collect the resulting basis vectors, we will succeed if eigenvectors corresponding to distinct eigenvalues are orthogonal. A wide class of matrices will produce this result.

5.7.6 Definitions and Comments. If A is a matrix with complex entries, the *conjugate transpose* of A, denoted by A^*, is formed by taking the complex conjugate of each element of A^t, or equivalently by conjugating every element of A and then transposing. For example, if

$$A = \begin{bmatrix} 1 & 2+i & 3-4i \\ 5+7i & 6-8i & 3 \\ 2-i & 4+9i & 10-5i \end{bmatrix}, \quad \text{then} \quad A^* = \begin{bmatrix} 1 & 5-7i & 2+i \\ 2-i & 6+8i & 4-9i \\ 3+4i & 3 & 10+5i \end{bmatrix}.$$

Exactly as in (5.1.4), we have

$$(A^*)^* = A, \quad (A+B)^* = A^* + B^*, \quad (cA)^* = \bar{c}A^*, \quad (AB)^* = B^*A^*.$$

The matrix A is *Hermitian* if $A = A^*$. Notice that if the entries of A are all real, then Hermitian is the same as symmetric: $A = A^t$. An example of a Hermitian matrix is

$$A = \begin{bmatrix} 1 & 2+i & 3-4i \\ 2-i & 6 & 3 \\ 3+4i & 3 & -10 \end{bmatrix}.$$

On the main diagonal we have $a_{ii} = \overline{a}_{ii}$, so that a_{ii} is real for all i.

The inner product of two vectors x and y in $\mathbb{C}^n$ can be expressed conveniently in terms of the conjugate transpose operation. If we regard x and y as column vectors, we have

$$\langle x, y \rangle = x_1 \overline{y}_1 + \cdots + x_n \overline{y}_n = y^* x. \tag{1}$$

If B is an n by n matrix, we have the useful identity

$$\langle Bx, y \rangle = \langle x, B^* y \rangle.$$

To see this, observe that by (1),

$$\langle x, B^* y \rangle = (B^* y)^* x = y^* B^{**} x = y^* B x = \langle Bx, y \rangle. \tag{2}$$

We may now prove some highly desirable properties of Hermitian matrices.

5.7.7 Theorem. *If A is Hermitian (in particular, if A is real and symmetric), then the eigenvalues of A are real, and eigenvectors corresponding to distinct eigenvalues are orthogonal.*

Proof. Let x be an eigenvector for the eigenvalue λ. Then by (2), $\langle \lambda x, x \rangle = \langle Ax, x \rangle = \langle x, A^* x \rangle = \langle x, Ax \rangle = \langle x, \lambda x \rangle$. Thus $\lambda \langle x, x \rangle = \overline{\lambda} \langle x, x \rangle$, and since $\langle x, x \rangle$ is real and greater than 0, we have $\lambda = \overline{\lambda}$, which says that λ is real.

Now let $Ax_1 = \lambda_1 x_1$ and $Ax_2 = \lambda_2 x_2$, where λ_1 and λ_2 are distinct eigenvalues. Then $\lambda_1 \langle x_1, x_2 \rangle = \langle \lambda_1 x_1, x_2 \rangle = \langle Ax_1, x_2 \rangle = \langle x_1, A^* x_2 \rangle = \langle x_1, Ax_2 \rangle = \langle x_1, \lambda_2 x_2 \rangle = \overline{\lambda_2} \langle x_1, x_2 \rangle = \lambda_2 \langle x_1, x_2 \rangle$ since λ_2 is real. But by hypothesis, $\lambda_1 \neq \lambda_2$, and it follows that $\langle x_1, x_2 \rangle = 0.$ ∎

In view of the discussion in (5.7.5), we know that if A is Hermitian and has distinct eigenvalues, then we can find an orthonormal basis consisting of eigenvectors of A. (The assumption of distinct eigenvalues is unnecessary, and will be removed in the next chapter.) Thus in the proof of (5.7.4), the change of basis matrix P has columns $x_1, \ldots, x_n$ which are mutually perpendicular unit vectors. Such a matrix is called *unitary*; if the entries of P are real, then P is called *orthogonal*.

5.7.8 Theorem. *If U is a square matrix, the following conditions are equivalent.*

(a) *U is unitary, that is, the columns are orthonormal vectors in $\mathbb{C}^n$.*

(b) *The rows of U are orthonormal vectors in $\mathbb{C}^n$.*

(c) *U is nonsingular and $U^{-1} = U^*$.*

If any, and hence all, of these conditions hold, then $|\det U| = 1$.

Proof. If the columns of U are $x_1, \ldots, x_n$, then the rows of U^* are $x_1^*, \ldots, x_n^*$. Thus the ij element of U^*U is $x_i^* x_j$, and it follows that orthonormality of the x_i is equivalent to $U^*U = I$. Thus (a) and (c) are equivalent. Similarly, orthonormality of the rows of U is equivalent to $UU^* = I$, and therefore (b) and (c) are equivalent. Finally, if $UU^* = I$, then

$$1 = \det I = (\det U)(\det U^*) = (\det U)(\det \overline{U})$$

since the determinant of the transpose of a matrix is the same as the determinant of the matrix itself (Section 5.2, Problem 5). Since the complex conjugate of a sum is the sum of the conjugates, and the conjugate of a product is the product of the conjugates, we have $\det \overline{U} = \overline{(\det U)}$, and therefore $1 = |\det U|^2$. Since absolute values are nonnegative, we have $|\det U| = 1$. ∎

If an inner product space has an orthonormal basis consisting of eigenvectors of a linear operator T (or a matrix A), we say that T (or A) can be *unitarily diagonalized*. Thus we have proved that any Hermitian matrix A can be unitarily diagonalized; there is a unitary matrix U such that $U^*AU = I$. If A is real and symmetric, then the eigenvalue problem can be solved without introducing complex numbers, the change of basis matrix U is orthogonal, and we have $U^t AU = I$. We say that A can be *orthogonally diagonalized*.

5.7.9 Quadratic Forms. Here is a major application of the theory of eigenvalues and eigenvectors. If A is a real symmetric matrix, the *quadratic form* associated with A is

$$x^t Ax = \sum_{i,j=1}^n a_{ij} x_i x_j.$$

Let L be an orthogonal matrix such that $L^t AL = D = \mathrm{diag}(\lambda_1, \ldots, \lambda_n)$. If we change variables by $x = Ly$, then

$$x^t Ax = y^t L^t ALy = y^t Dy = \sum_{i=1}^n \lambda_i y_i^2.$$

This process is called reducing a quadratic form to a sum of squares. The real symmetric matrix A is said to be *nonnegative definite* if the quadratic form $x^t Ax$ is always nonnegative, regardless of the values of the variables x_i. Equivalently, $\sum_{i=1}^n \lambda_i y_i^2 \geq 0$ for all values of the y_i. Thus

A is nonnegative definite if and only if all eigenvalues of A are nonnegative.

A is said to be *positive definite* if the quadratic form is strictly positive except when all $x_i = 0$ (in which case the quadratic form is 0). Thus

A is positive definite if and only if all eigenvalues of A are strictly positive.

As an example, let q be the quadratic form

$$3x^2 + 2xy + 3y^2 = [x \; y] \begin{bmatrix} 3 & 1 \\ 1 & 3 \end{bmatrix} \begin{bmatrix} x \\ y \end{bmatrix},$$

with associated matrix $A = \begin{bmatrix} 3 & 1 \\ 1 & 3 \end{bmatrix}$. The determinant of $A - \lambda I$ is

$$\begin{vmatrix} 3-\lambda & 1 \\ 1 & 3-\lambda \end{vmatrix} = \lambda^2 - 6\lambda + 8,$$

and there are two eigenvalues $\lambda = 2$ and $\lambda = 4$.

Let u be a column vector with components x and y. When $\lambda = 2$, the equations $(A - \lambda I)u = 0$ reduce to $x + y = 0$, so $(1, -1)$ is an eigenvector; normalize it to get $(1/\sqrt{2}, -1/\sqrt{2})$.

When $\lambda = 4$, the equations $(A - \lambda I)u = 0$ become $-x + y = 0$, so the normalized eigenvector is $(1/\sqrt{2}, 1/\sqrt{2})$. Thus the change of basis matrix is

$$L = \begin{bmatrix} 1/\sqrt{2} & 1/\sqrt{2} \\ -1/\sqrt{2} & 1/\sqrt{2} \end{bmatrix},$$

and $L^t A L$ does turn out to be $D = \begin{bmatrix} 2 & 0 \\ 0 & 4 \end{bmatrix}$ as expected. If $\begin{bmatrix} x \\ y \end{bmatrix} = L \begin{bmatrix} x' \\ y' \end{bmatrix}$, that is,

$$x = \frac{1}{\sqrt{2}}x' + \frac{1}{\sqrt{2}}y' \quad \text{and} \quad y = -\frac{1}{\sqrt{2}}x' + \frac{1}{\sqrt{2}}y',$$

then

$$q = 3\left(\frac{(x')^2}{2} + \frac{(y')^2}{2} + x'y'\right) + 2\left(\frac{-(x')^2}{2} + \frac{(y')^2}{2}\right) + 3\left(\frac{(x')^2}{2} + \frac{(y')^2}{2} - x'y'\right)$$

$$= 2(x')^2 + 4(y')^2 = [x'\ y']D\begin{bmatrix} x' \\ y' \end{bmatrix},$$

as expected.

Problems For Section 5.7

1. Show that the product of unitary matrices is unitary. What about the sum?

2. If λ is an eigenvalue of T, show that λ^k is an eigenvalue of T^k for every positive integer k.

3. Find the eigenvalues and eigenvectors of the matrix

$$A = \begin{bmatrix} 2 & 1 & 0 \\ 0 & 2 & 0 \\ 0 & 0 & 1 \end{bmatrix},$$

and conclude that A is not diagonalizable.

4. Show that a square matrix is invertible if and only if 0 is not an eigenvalue of A.

5. Suppose that A is a 2 by 2 Hermitian matrix with eigenvalues $\lambda = 3$ and $\lambda = 5$. If $(2,4)$ is an eigenvector for $\lambda = 3$ and $(-7, y)$ (with y real) is an eigenvector for $\lambda = 5$, find y.

6. If $U^* A U = D$, where U is unitary and D is a diagonal matrix with all entries real, show that A must be Hermitian.

7. Show that if A is diagonalizable, then the determinant of A is the product of the eigenvalues of A, counting multiplicity. (This holds even if A is not diagonalizable, as we will see in Section 6.1, Problem 4.)

8. If A is similar to the matrix $D = \operatorname{diag}(\lambda_1, \ldots, \lambda_n)$, with $A = PDP^{-1}$, exploit this fact to give an efficient method for computing powers of A.

9. Here is another method to reduce a quadratic form to a sum of squares. As an example, consider $q = 3x^2 + 2xy - y^2 = 3(x^2 + \frac{2}{3}xy + (\quad)) - y^2 - (\quad)$. Fill in the blanks by completing the square, and continue the process to express q as $aX^2 + bY^2$ for appropriate a and b. Identify the basis-changing matrix P and find a basis for the new coordinate system.

10. Let $q = 3x^2 - 6y^2 + z^2 + 6xy + 18xz$. Indicate how to reduce q to a sum of squares by completing the square. Carry out the first few steps so that the general pattern will be clear.

Sylvester's Law of Inertia states that no matter how a quadratic form is reduced to a sum of squares $a_1 x_1^2 + \cdots + a_r x_r^2$, the number of positive a_i and the number of negative a_i are invariant.

11. Show that a unitary matrix U preserves norms, that is, $||Ux|| = ||x||$ for all x.

12. Show that if λ is an eigenvalue of the unitary matrix U, then $|\lambda| = 1$.

6

Theory of Linear Operators

6.1 Jordan Canonical Form

Let T be a linear operator on the finite-dimensional vector space V over the complex field $\mathbb{C}$. We ask the following basic questions: When can T be diagonalized? When can T be unitarily diagonalized? If T cannot be diagonalized, can we find a matrix representing T that is close in some sense to diagonal form?

We will begin with the third question, which turns out to give considerable insight into the other two. We will prove eventually that any square matrix A with complex entries is similar (see 5.5.7) to a matrix in *Jordan canonical form*. The Jordan canonical form is a direct sum of *Jordan blocks,* which are matrices with the same entry λ (an eigenvalue of A) down the main diagonal, and with the same entry 1 down the diagonal directly to the northeast of the main diagonal. Here is an example with 7 Jordan blocks associated with the eigenvalue $\lambda = 3$. (In general, for each eigenvalue λ_i of A, there will be Jordan blocks corresponding to λ_i.)

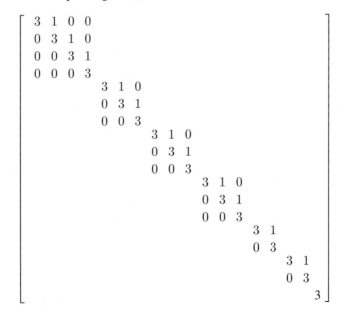

We actually have an 18 by 18 matrix; the missing entries are taken to be 0. Notice that the last block, which is 1 by 1, is legal; the Jordan condition is satisfied vacuously since there is no northeast diagonal in this case.

Let's look at the first Jordan block

$$J_1 = \begin{bmatrix} 3 & 1 & 0 & 0 \\ 0 & 3 & 1 & 0 \\ 0 & 0 & 3 & 1 \\ 0 & 0 & 0 & 3 \end{bmatrix}$$

and do some computation. With $\lambda = 3$ we have

$$J_1 - \lambda I = \begin{bmatrix} 0 & 1 & 0 & 0 \\ 0 & 0 & 1 & 0 \\ 0 & 0 & 0 & 1 \\ 0 & 0 & 0 & 0 \end{bmatrix}$$

and

$$(J_1 - \lambda I)^2 = \begin{bmatrix} 0 & 0 & 1 & 0 \\ 0 & 0 & 0 & 1 \\ 0 & 0 & 0 & 0 \\ 0 & 0 & 0 & 0 \end{bmatrix}.$$

Thus squaring $J_1 - \lambda I$ causes a migration of 1's to the adjacent northeast diagonal, as you can verify. Similarly,

$$(J_1 - \lambda I)^3 = \begin{bmatrix} 0 & 0 & 0 & 1 \\ 0 & 0 & 0 & 0 \\ 0 & 0 & 0 & 0 \\ 0 & 0 & 0 & 0 \end{bmatrix}$$

and $(J_1 - \lambda I)^4 = 0$.

If J_2 is the second Jordan block, we have

$$J_2 - \lambda I = \begin{bmatrix} 0 & 1 & 0 \\ 0 & 0 & 1 \\ 0 & 0 & 0 \end{bmatrix}, \qquad (J_2 - \lambda I)^2 = \begin{bmatrix} 0 & 0 & 1 \\ 0 & 0 & 0 \\ 0 & 0 & 0 \end{bmatrix}, \qquad (J_2 - \lambda I)^3 = 0;$$

the same results hold for J_3 and J_4, which coincide with J_2. For J_5 and J_6 ($= J_5$) we have $J_5 - \lambda I = \begin{bmatrix} 0 & 1 \\ 0 & 0 \end{bmatrix}$ and $(J_5 - \lambda I)^2 = 0$. Finally for the 1 by 1 block J_7 we have $J_7 - \lambda I = 0$.

If $J = J(\lambda)$ is the entire Jordan matrix corresponding to the eigenvalue λ, the block structure of J can be recovered from the ranks of the various powers of $J - \lambda I$. For example, suppose that we are presented with the following data, for an 18×18 matrix J:

$$\text{rank}(J - \lambda I)^4 = 0, \quad \text{rank}(J - \lambda I)^3 = 1, \quad \text{rank}(J - \lambda I)^2 = 5, \quad \text{rank}(J - \lambda I) = 11.$$

Then $(J - \lambda I)^4 = 0$ but $(J - \lambda I)^3 \neq 0$, and from this we can deduce that the largest Jordan block has order 4 (i.e., is a 4 by 4 matrix). If J_i is a block of order less than 4, then $(J_i - \lambda I)^3 = 0$, and since the rank of a direct sum of matrices is the sum of the individual ranks (see (5.4.2 (e)), the rank of $(J - \lambda I)^3$ is the number of blocks of order 4. Thus in this case, there is exactly one block of order 4. Similarly,

$$\text{rank}(J - \lambda I)^2 = 2 \text{ (number of blocks of order 4) } + 1 \text{ (number of blocks of order 3)},$$

which yields $5 = 2(1) +$ (number of blocks of order 3). Thus there are exactly 3 blocks of order 3. (Notice the pattern: each block of order 4 contributes $4 - 2 = 2$ to the rank of $(J - \lambda I)^2$, and each block of order 3 contributes $3 - 2 = 1$.)

Now, using # as an abbreviation for number,

$$\text{rank}(J - \lambda I) = 3 \text{ (\# of blocks of order 4) } + 2 \text{ (\# of blocks of order 3)}$$

$$+ 1 \text{ (\# of blocks of order 2)},$$

so

$$11 = 3(1) + 2(3) + 1 \text{ (\# of blocks of order 2)},$$

and we conclude that there are exactly 2 blocks of order 2. Finally, the number of blocks of order 1 is

$$\text{the order of } J - 4 \text{ (\# of blocks of order 4) } - 3 \text{ (\# of blocks of order 3)}$$

$$- 2 \text{ (\# of blocks of order 2)}$$

$$= 18 - 4(1) - 3(3) - 2(2) = 1,$$

so there is exactly 1 block of order 1.

The above discussion outlines a systematic method for recovering the Jordan matrix J from the ranks of the powers of $(J - \lambda I)$. Since the rank of a matrix is the dimension of the image of the associated linear operator T (see (5.4.5) and (5.5.5)), it follows that the number of Jordan blocks of a given order is completely determined by T. We have the following important conclusion.

6.1.1 Uniqueness of Jordan Canonical Form. The Jordan canonical form of a given linear operator is unique, up to a permutation of the individual blocks. Thus we cannot distinguish between $J_1 \oplus J_2 \oplus J_3 \oplus J_4$ and $J_2 \oplus J_4 \oplus J_3 \oplus J_1$. What we are doing here is permuting basis vectors. For example, if x_1, x_2, x_3 and x_4 form a basis for V and we list the vectors as x_2, x_4, x_3, x_1, this slight adjustment is technically a basis change, and we move from an old matrix A representing the given linear transformation T, to a new matrix B. But B is obtained from A by simply permuting columns.

6.1.2 Generalized Eigenvectors. Let's again focus our attention on the matrix J_1. If we wish to study J_1 in isolation, we are concentrating on the subspace V_1 consisting of vectors whose components are 0 except possibly in the first 4 places. Rather than work with the original linear operator T, we are restricting T to V_1 to produce another linear operator T_1 on V_1. (This is legal because if $x \in V_1$, then $Tx \in V_1$; if a vector x has 0's except in positions 1–4, the same is true of Tx, because J is the direct sum of the blocks J_i.) Now J_1 represents T_1 with respect to some basis x_1, x_2, x_3, x_4 for V_1, and from the form of J_1 we have

$$Tx_1 = \lambda x_1, \quad Tx_2 = x_1 + \lambda x_2, \quad Tx_3 = x_2 + \lambda x_3, \quad Tx_4 = x_3 + \lambda x_3, \qquad (1)$$

where $\lambda = 3$ and $T = T_1$ on V_1. Thus

$$(T - \lambda I)x_1 = 0, \quad (T - \lambda I)x_2 = x_1, \quad (T - \lambda I)x_3 = x_2, \quad (T - \lambda I)x_4 = x_3, \quad (2)$$

and consequently

$$(T - \lambda I)x_1 = 0, \quad (T - \lambda I)^2 x_2 = (T - \lambda I)x_1 = 0,$$
$$(T - \lambda I)^3 x_3 = (T - \lambda I)^2 x_2 = 0, \quad (T - \lambda I)^4 x_4 = (T - \lambda I)^3 x_3 = 0. \tag{3}$$

Thus among the basis vectors x_1, x_2, x_3, x_4, only x_1 is an eigenvector. (For example, if $(T - \lambda I)x_2 = 0$, then $x_1 = 0$ by (2), a contradiction.) In assembling a basis for the eigenspace of T, each block contributes only a single eigenvector. But all the x_i are *generalized eigenvectors*; that is, $(T - \lambda I)^k x_i = 0$ for some positive integer k. The Jordan canonical form does not produce a basis of eigenvectors, but it comes close:

(4)   If T is a linear operator on V, then there is a basis for V consisting of generalized eigenvectors.

Notice that because the Jordan canonical form is a direct sum of Jordan blocks, we can work with each eigenvalue individually and assemble the results at the end. More precisely,

(5)   If $y_1, \ldots, y_p$ are generalized eigenvectors corresponding to distinct eigenvalues $\lambda_1, \ldots, \lambda_p$, then $y_1, \ldots, y_p$ are linearly independent.

This follows because if W is the direct sum of subspaces W_i, $i = 1, \ldots, p$, and $y_i \in W_i$, $i = 1, \ldots, p$, then the y_i are linearly independent. For if $a_1 y_1 + \cdots + a_p y_p = 0$, then $a_1 y_1 + \cdots + a_p y_p = 0 y_1 + \cdots + 0 y_p$, and therefore $a_i = 0$ for all i.

Problems For Section 6.1

1. Suppose that $J = J(\lambda)$ is a 14 by 14 Jordan matrix corresponding to the eigenvalue λ. If

$$\text{rank}(J - \lambda I)^3 = 0, \quad \text{rank}(J - \lambda I)^2 = 2, \quad \text{and} \quad \text{rank}(J - \lambda I) = 7,$$

determine the number of Jordan blocks of each order.

2. Let J be a 3 by 3 Jordan matrix corresponding to the eigenvalue λ. Under what conditions will J consist of a single Jordan block of order 3?

3. In Problem 2, under what conditions will J consist of 3 Jordan blocks of order 1? Note that in this case, J is of the form

$$\begin{bmatrix} \lambda & 0 & 0 \\ 0 & \lambda & 0 \\ 0 & 0 & \lambda \end{bmatrix}.$$

4. Show that if A is any square matrix, then the determinant of A is the product of the eigenvalues of A, counting multiplicity.

6.2 The Minimal and Characteristic Polynomials

If A is a square matrix with complex entries, we can compute polynomials in A, that is, matrices of the form

$$p(A) = a_m A^m + \cdots + a_1 A + a_0 I;$$

if A represents the linear operator T, then $p(A)$ represents $p(T)$. We ask the following question:

For which nonzero polynomials p is $p(A)$ the zero matrix? (If $p(A) = 0$, we say that p is *satisfied* by A.)

It is not obvious that there are any polynomials at all that satisfy this condition, so let's deal with this problem first. If A is n by n, then A has n^2 entries, and we can think of A as a vector with n^2 components. The collection of all n by n matrices is an n^2-dimensional vector space, and a basis for this space (analogous to the standard basis for $\mathbb{C}^{n^2}$) is formed by matrices E_{ij} having an entry 1 in row i and column j, and zeros elsewhere. The matrices $I, A, A^2, \ldots, A^{n^2}$ form a set of $n^2 + 1$ vectors in an n^2-dimensional space, and therefore are linearly dependent by (5.3.3g). Thus there are scalars $a_0, a_1, \ldots, a^{n^2}$, not all zero, such that $a_0 I + a_1 A + \cdots + a_{n^2} A^{n^2} = 0$, as desired.

Now we might suspect that we can find polynomials p of lower degree than n^2 satisfied by A, and this turns out to be correct. What is the nonzero polynomial of lowest degree satisfied by A? This question is a bit ambiguous, because if $p(A) = 0$, then $2p(A) = 0$, $3p(A) = 0$, in fact $kp(A) = 0$ for any constant k. We can remove the ambiguity by restricting to *monic* polynomials. A monic polynomial looks like

$$x^m + a_{m-1} x^{m-1} + \cdots + a_1 x + a_0;$$

the *leading coefficient* (the coefficient of the highest power of x) is 1. If p and q are monic polynomials of lowest degree satisfied by A, then $p = q$. For if p and q are not identical, then $p - q$ is a nonzero polynomial of lower degree than that of p and q, and $p(A) - q(A) = 0$. By dividing $p - q$ by its leading coefficient we can produce a monic polynomial of lower degree satisfied by A, which is a contradiction. The following definition is therefore sensible.

6.2.1 Definition. The *minimal polynomial* of a square matrix A is the monic polynomial p of lowest degree such that $p(A) = 0$. The minimal polynomial of A will be denoted by $m_A(x)$, or simply by $m(x)$ if it is clear which matrix A we are talking about.

The minimal polynomial has the following basic property.

6.2.2 Theorem. *Let A be a square matrix with minimal polynomial $m(x)$, and let $f(x)$ be an arbitrary polynomial. Then $f(A) = 0$ if and only if $f(x)$ is a multiple of $m(x)$; that is, there is a polynomial $g(x)$ such that $f(x) = g(x)m(x)$.*

Proof. If $f(x) = g(x)m(x)$, then $f(A) = g(A)m(A) = g(A)0 = 0$, so assume that $f(A) = 0$. Divide $f(x)$ by $m(x)$ to produce a quotient $q(x)$ and remainder $r(x)$:

$$f(x) = q(x)m(x) + r(x),$$

where $r(x)$ has lower degree than $m(x)$. (To avoid having to consider the case $r(x) = 0$ separately, it is convenient to regard the zero polynomial as having degree $-\infty$.) Now $f(A) = m(A) = 0$, and therefore $r(A) = 0$. But if $r(x) \neq 0$, then by dividing $r(x)$ by its leading coefficient we produce a monic polynomial of lower degree than $m(x)$ that is satisfied by A, a contradiction. Therefore $r(x) = 0$, and $f(x) = q(x)m(x)$. ∎

Minimal polynomials of Jordan blocks can be computed easily; let's look at an example:

$$J = \begin{bmatrix} 3 & 1 & 0 & 0 \\ 0 & 3 & 1 & 0 \\ 0 & 0 & 3 & 1 \\ 0 & 0 & 0 & 3 \end{bmatrix},$$

a block of order 4 for the eigenvalue $\lambda = 3$. In Section 6.1 we found that $(J - \lambda I)^4 = 0$, but $(J - \lambda I)^k \neq 0$ for $k < 4$. By (6.2.2), the minimal polynomial $m(x)$ of J is a divisor of $(x - \lambda)^4$, so it must be a power of $x - \lambda$. But since no power of $x - \lambda$ lower than the fourth can be satisfied by J, we must have

$$m(x) = (x - \lambda)^4.$$

The analysis is the same regardless of the particular eigenvalue λ or the size of the Jordan block.

6.2.3 The Minimal Polynomial of a Jordan Block. If J is a Jordan block of order k associated with the eigenvalue λ, then the minimal polynomial of J is

$$m_J(x) = (x - \lambda)^k.$$

Let's return to the computation at the beginning of Section 6.1. We had an 18 by 18 Jordan matrix J with one block J_1 of order 4, three blocks J_2, J_3, and J_4 of order 3, two blocks J_5 and J_6 of order 2, and one block J_7 of order 1. If $m_i(x)$ is the minimal polynomial of J_i, we have just seen that

$$m_1(x) = (x - \lambda)^4, \quad m_2(x) = m_3(x) = m_4(x) = (x - \lambda)^3,$$

$$m_5(x) = m_6(x) = (x - \lambda)^2, \quad m_7(x) = x - \lambda.$$

Since J is the direct sum of the blocks J_i, we can compute a polynomial $p(J)$ by working on each block separately; in other words, $p(J)$ is the direct sum of the $p(J_i)$. But as the above computation suggests, any polynomial satisfied by J_1 will automatically be satisfied by the other blocks. Thus:

The minimal polynomial of $J = J(\lambda)$ is $(x - \lambda)^r$, where r is the size of the largest Jordan block corresponding to λ.

We may now prove a general result about minimal polynomials.

6.2.4 Theorem. *Let A be an n by n matrix with eigenvalues $\lambda_1, \ldots, \lambda_k$ and Jordan canonical form $J = J(\lambda_1) \oplus \cdots \oplus J(\lambda_k)$. (The λ_i are distinct, but k may be less than n; in other words, some of the λ_i may be multiple roots of the equation $\det(A - \lambda I) = 0$.) If r_i*

is the size of the largest Jordan block corresponding to λ_i, *then the minimal polynomial of* A *is*

$$m_A(x) = (x - \lambda_1)^{r_1} \cdots (x - \lambda_k)^{r_k}.$$

Proof. We have just seen that the minimal polynomial of $J(\lambda_i)$ is $(x - \lambda_i)^{r_i}$, $i = 1, \ldots, k$. If $f(x)$ is any polynomial satisfied by J, we must have $f(J(\lambda_i)) = 0$, so by (6.2.2), $f(x)$ is a multiple of each factor $(x - \lambda_i)^{r_i}$, and consequently $f(x)$ is a multiple of $g(x) = (x - \lambda_1)^{r_1} \cdots (x - \lambda_k)^{r_k}$. But $g(J(\lambda_i)) = 0$ for all i, and since J is the direct sum of the $J(\lambda_i)$, we have $g(J) = 0$. Therefore the required minimal polynomial is $g(x)$.∎

We may now give a criterion for diagonalizability of a matrix.

6.2.5 Theorem. *The matrix A is diagonalizable if and only if $m_A(x)$ is a product of distinct linear factors.*

Proof. Diagonalizability of A means that all Jordan blocks are of order 1. By (6.2.4), this is equivalent to saying that

$$m_A(x) = (x - \lambda_1) \cdots (x - \lambda_k),$$

a product of distinct linear factors. ∎

It is still not clear how to produce, for a given matrix A, an explicit nonzero polynomial satisfied by A. We now address this question.

6.2.6 Definitions and Comments. The *characteristic polynomial* of the n by n matrix A is

$$c_A(x) = \det(xI - A).$$

(We simply write $c(x)$ if it is clear which A is intended.) If we multiply each entry of $A - xI$ by -1 or, equivalently, multiply each row (or each column) by -1, we multiply the determinant by $(-1)^n$. Therefore

$$c_A(x) = (-1)^n \det(A - xI).$$

The equation $\det(A - xI) = 0$ (equivalently, $\det(xI - A) = 0$) is called the *characteristic equation* of A; its roots are the eigenvalues of A.

Now in the expansion of the determinant of $xI - A$, there will be only one term of degree n, namely x^n. This term is generated by the n occurrences of x on the main diagonal of A. Thus $c_A(x)$ is a monic polynomial, and consequently

$$c_A(x) = (x - \lambda_1)^{s_1} \cdots (x - \lambda_k)^{s_k}$$

where the λ_i are the eigenvalues of A and s_i is the multiplicity of λ_i.

This looks very similar to the expression for $m_A(x)$ in (6.2.4), and in fact we will see in the proof of the next theorem that $s_i \geq r_i$ for all $i = 1, \ldots, k$.

6.2.7 Cayley-Hamilton Theorem.   *Every square matrix satisfies its characteristic equation. In other words, if A is n by n with characteristic polynomial $c_A(x)$, then $c_A(A) = 0$. In fact $c_A(x)$ is a (polynomial) multiple of the minimal polynomial $m_A(x)$.*

Proof.   Let $J = J(\lambda_1) \oplus \cdots \oplus J(\lambda_k)$ be the Jordan canonical form of A. Then $\det(xI - A) = \det(xI - J)$ (see (5.7.2)), and (because of the direct sum) this determinant is the product of $\det(xI - J(\lambda_i))$, $i = 1, \dots, k$. But

$$\det(xI - J(\lambda_i)) = (x - \lambda_i)^{t_i},$$

where t_i is the size of the Jordan matrix $J(\lambda_i)$. (To verify this, look again at the Jordan matrix analyzed at the beginning of Section 6.1.) Therefore

$$c_A(x) = (x - \lambda_1)^{t_1} \cdots (x - \lambda_k)^{t_k}.$$

(In fact $t_i = s_i$, the multiplicity of λ_i; see (6.2.6).) But the size t_i of the entire Jordan matrix $J(\lambda_i)$ is certainly as least as big as the size r_i of the largest Jordan block within $J(\lambda_i)$. Thus by (6.2.4), $c_A(x)$ is a multiple of $m_A(x)$, and it follows that $c_A(A) = 0$. ∎

6.2.8 Example.   Let

$$A = \begin{bmatrix} 0 & 14 & 1 \\ 0 & 5 & 0 \\ -4 & 13 & 4 \end{bmatrix};$$

find the Jordan canonical form of A and its minimal and characteristic polynomials. Find a nonsingular matrix P such that $P^{-1}AP$ is the Jordan canonical form of A.

The determinant of $xI - A$ is $x^3 - 9x^2 + 24x - 20 = (x - 2)^2(x - 5)$, so the eigenvalues of A are $\lambda = 2$ (with multiplicity 2) and $\lambda = 5$. An eigenvector for $\lambda = 5$ is $(3, 1, 1)$, but we can find only one linearly independent eigenvector for $\lambda = 2$, e.g., $x_1 = (1, 0, 2)$. (The equation $(A - 2I)(a, b, c)^t = 0$ reduces to $b = 0$, $c = 2a$.) Following the analysis in (6.1.2), we look for a vector x_2 such that $(A - 2I)x_2 = x_1$. Thus we must solve the equation

$$\begin{bmatrix} -2 & 14 & 1 \\ 0 & 3 & 0 \\ -4 & 13 & 2 \end{bmatrix} \begin{bmatrix} a \\ b \\ c \end{bmatrix} = \begin{bmatrix} 1 \\ 0 \\ 2 \end{bmatrix},$$

which reduces to $b = 0$, $c = 2a + 1$. Therefore we may take $x_2 = (1, 0, 3)$. Since a Jordan block of order k corresponds to a chain of k generalized eigenvectors (see (1) of (6.1.2)), it follows that the Jordan canonical form of A has one block of order 2 and one block of order 1. The Jordan canonical form is

$$J = \begin{bmatrix} 2 & 1 & 0 \\ 0 & 2 & 0 \\ 0 & 0 & 5 \end{bmatrix}.$$

By (6.2.4), the minimal polynomial of A is $(x - 2)^2(x - 5)$, which coincides with the characteristic polynomial. The vectors x_1, x_2, and $x_3 = (3, 1, 1)$ form a basis of generalized eigenvectors, with

$$(A - 2I)x_1 = 0, \quad (A - 2I)x_2 = x_1, \quad (A - 5I)x_3 = 0.$$

Thus we may take P to have columns x_1, x_2, x_3, so that

$$P = \begin{bmatrix} 1 & 1 & 3 \\ 0 & 0 & 1 \\ 2 & 3 & 1 \end{bmatrix}.$$

Problems For Section 6.2

1. Let J be a Jordan block of size r corresponding to λ, so that J has λ's down the main diagonal, 1's down the adjacent northeast diagonal, and 0's elsewhere. Show that the minimal and characteristic polynomials of J coincide.

2. If A is an $n \times n$ matrix with distinct eigenvalues $\lambda_1, \ldots, \lambda_n$, write down the characteristic and minimal polynomials of A, and the Jordan canonical form.

3. Let A be a 3 by 3 matrix with characteristic polynomial $c(x) = (x - \lambda_1)^2 (x - \lambda_2)$. Write down the possible minimal polynomials, and in each case give the associated Jordan canonical form. (By (6.1.1), the Jordan form is determined up to a permutation of the individual blocks.)

4. Let A be a 3 by 3 matrix with characteristic polynomial $c(x) = (x - \lambda)^3$. Write down the possible minimal polynomials, and in each case give the associated Jordan canonical form.

5. Problems 2, 3 and 4 show that if A is 3 by 3 (or smaller), the Jordan canonical form is determined by the characteristic and minimal polynomials. Give an example to show that this property fails when A is 4 by 4 (or larger).

6. Show how the Cayley-Hamilton Theorem can be used to compute the inverse of a matrix.

6.3 The Adjoint of a Linear Operator

Theorem 6.2.5 gives a necessary and sufficient condition for a matrix A to be diagonalizable, and we now look for conditions under which A can be unitarily diagonalized. Suppose that V is an n-dimensional inner product space, and we define $g(x)$ to be the inner product $\langle x, y \rangle$ where y is a fixed element of V. Then by the definition of an inner product (see (5.6.1)), g is a linear function from V to the underlying field (the complex numbers); such a mapping is called a *linear functional*. In fact every linear functional is of this form.

6.3.1 Theorem. *If g is a linear functional on the n-dimensional inner product space V, there is a unique element y in V such that $g(x) = \langle x, y \rangle$ for every $x \in V$. Specifically, if $x_1, \ldots, x_n$ form an orthonormal basis for V, then*

$$y = \sum_{i=1}^{n} \overline{g(x_i)} x_i. \tag{1}$$

Proof. If y is given by (1) and $f(x) = \langle x, y \rangle$, we must show that $f(x) = g(x)$ for all x, and since the x_j form a basis, it suffices to show that $f(x_j) = g(x_j)$ for all j. But

$$f(x_j) = \langle x_j, y \rangle = \left\langle x_j, \sum_{i=1}^{n} \overline{g(x_i)} x_i \right\rangle.$$

By orthonormality of the x_i and the fact that an inner product is conjugate linear in the second variable (see (4) of (5.6.1)), we have $f(x_j) = g(x_j)$. ∎

In Section 5.7, we discussed the conjugate transpose A^* of a matrix A, but we omitted an interesting question. If T is a linear operator represented by A, what can we say about the linear operator represented by A^*? We will approach this question by giving an abstract definition of the adjoint T^* of a linear operator T, and then proving that T^* is represented by A^*.

6.3.2 Definitions and Comments. Let T be a linear operator on the n-dimensional inner product space V. If $y \in V$, let g be the linear functional given by

$$g(x) = \langle Tx, y \rangle.$$

By (6.3.1), there is a unique element $y_0 \in V$ such that $g(x) = \langle x, y_0 \rangle$ for all $x \in V$. If we change y, then y_0 may change as well, but since y_0 is uniquely determined by y, we can express y_0 as a function of y, and we write

$$y_0 = T^* y.$$

Thus T^*, which is called the *adjoint* of T, is defined by the requirement that

$$\langle Tx, y \rangle = \langle x, T^* y \rangle \quad \text{for all } x, y \in V.$$

(This terminology is completely unrelated to the adjoint of a matrix as defined in (5.2.3).)

The uniqueness statement in (6.3.1) implies that T^* is well defined, in other words, if $S : V \to V$ and $\langle Tx, y \rangle = \langle x, Sy \rangle$ for all $x, y \in V$, then $S = T^*$. If this is not convincing, observe that $\langle x, Sy - T^* y \rangle = 0$ for all $x, y \in V$, so that $Sy - T^* y$ is orthogonal to everything in V, and therefore orthogonal to itself. But a vector perpendicular to itself must be 0, and the result follows.

The adjoint T^* is a function from V to V, but we can't immediately jump to the conclusion that T^* is linear; this must be proved.

6.3.3 Theorem. *The adjoint T^* is a linear operator on V.*

Proof. Let x, y, and z be arbitrary elements of V. Then, using the definition of T^* and basic properties of the inner product, we have

$$\langle x, T^*(ay + bz) \rangle = \langle Tx, ay + bz \rangle$$
$$= \overline{a} \langle Tx, y \rangle + \overline{b} \langle Tx, z \rangle$$
$$= \overline{a} \langle x, T^* y \rangle + \overline{b} \langle x, T^* z \rangle$$
$$= \langle x, aT^* y + bT^* z \rangle.$$

Since x is arbitrary, we must have $T^*(ay + bz) = aT^*y + bT^*z$. ∎

Matrices provide a concrete interpretation of the adjoint, and establish a link to our earlier discussion of the conjugate transpose of a matrix (see (5.7.6)).

6.3.4 Theorem. *If the linear operator T is represented by the matrix A with respect to the orthonormal basis $\{x_1, \ldots, x_n\}$, then with respect to this basis, T^* is represented by the conjugate transpose A^* of A.*

Proof. By (5.6.6), $Tx_j = \sum_{i=1}^n \langle Tx_j, x_i \rangle x_i$. But $Tx_j = \sum_{i=1}^n a_{ij} x_i$ (see Equation (α) of (5.5.4)), so $a_{ij} = \langle Tx_j, x_i \rangle$. Similarly, if B is the matrix that represents T^*, then $b_{ij} = \langle T^*x_j, x_i \rangle$. Therefore

$$b_{ij} = \overline{\langle x_i, T^*x_j \rangle} = \overline{\langle Tx_i, x_j \rangle} = \overline{a_{ji}},$$

so $B = A^*$. ∎

The following properties of the adjoint are established quickly using the definition of T^*.

6.3.5 Theorem. *Let T and S be linear operators on the n-dimensional inner product space V. Then*
(1) $(T + S)^* = T^* + S^*$.
(2) $(cT)^* = \overline{c}T^*$ *for any complex number c, where cT is the linear operator that maps the element x to cTx.*
(3) $(TS)^* = S^*T^*$, *where TS is the composition $T \circ S$, and $S^*T^* = S^* \circ T^*$.*
(4) $T^{**} = T$.
(5) *If I is the identity operator, then $I^* = I$.*

Proof. This will make a good exercise (Problems 1-5). ∎

We can ask when a linear operator coincides with its adjoint, and thereby make a connection with Hermitian matrices.

6.3.6 Definition. The linear operator T is said to be *self-adjoint* if $T = T^*$. In view of (6.3.4), self-adjoint operators are those that are represented by matrices A satisfying $A^* = A$, that is, Hermitian matrices.

In the next section, we will identify precisely the class of linear operators that can be unitarily diagonalized. We will see that self-adjoint operators belong to this class but do not constitute the entire class.

Recall that a **projection operator** P maps the vector x to its projection on a subspace S; we call P the *projection operator onto S*. We proved that P is linear in the discussion following (5.6.9). The following result gives the essential properties of such operators.

6.3.7 Theorem. *Let P be the projection operator onto a subspace S. Then*
(a) *P is idempotent, that is, $P^2 = P$. Furthermore,*

(b) P is self-adjoint.

Conversely, if P is a self-adjoint idempotent operator on V and the image of P is the subspace S, then P is the projection operator onto S.

Proof. Any vector in S is projected onto itself, in particular $P(Px) = Px$, proving (a). Now for any x, the Projection Theorem (5.6.9) shows that $x - Px$ is orthogonal to S. But S is the image of P, so that for every x and y we have

$$\langle x - Px, Py \rangle = 0, \tag{1}$$

and consequently

$$\langle x, Py \rangle = \langle Px, Py \rangle = \langle x, P^*Py \rangle. \tag{2}$$

Thus $Py - P^*Py$ is orthogonal to every $x \in V$, in particular orthogonal to itself. It follows that $Py - P^*Py = 0$. But y is arbitrary, so $P = P^*P$. Take the adjoint of both sides to obtain $P^* = (P^*P)^* = P^*P^{**} = P^*P$ by (6.3.5) parts (3) and (4). But then $P = P^*$, proving (b).

Conversely, if P is idempotent and self-adjoint, then $P = PP = P^*P$, so that (2) holds for every x and y. But then (1) holds as well, which means that $z = x - Px$ is orthogonal to the image of P, namely S. By the uniqueness part of the Projection Theorem, Px must be the projection of x on S. ∎

Problems For Section 6.3

1–5. Prove statements (1) – (5) in (6.3.5).

6. Let T be a linear operator on the n-dimensional inner product space V, and let x be an element of V. Show that x belongs to the kernel of T (that is, $Tx = 0$) if and only if x is orthogonal to every element of the image of T^* (that is, $\langle x, T^*y \rangle = 0$ for every $y \in V$). We say that the subspaces $\ker T$ and $\operatorname{im} T^*$ are *orthogonal complements*.

7. Continuing Problem 6, show that the image of T and the kernel of T^* are orthogonal complements.

6.4 Normal Operators

We are going to show that a linear operator T on the n-dimensional inner product space V (over the complex field $\mathbb{C}$) can be unitarily diagonalized if and only if T is normal. Let's first define normal operators.

6.4.1 Definitions. Let V be an n-dimensional inner product space, and let T be a linear operator on V. We say that T is *normal* if T commutes with its adjoint, that is,

$$TT^* = T^*T.$$

If the linear operator T is represented by the matrix A with respect to a given basis, then normality of T translates to

$$AA^* = A^*A;$$

matrices having this property are referred to as normal also.

We have already seen many examples of normal operators. If T is self-adjoint, that is, $T = T^*$ (or in terms of matrices, $A = A^*$, the conjugate transpose of A), then $TT^* = T^*T = T^2$, so that T is normal. Here is another major example:

T is said to be *unitary* if T is bijective and the inverse T^{-1} coincides with T^*. In terms of matrices, A is invertible and $A^{-1} = A^*$, the conjugate transpose of A. Thus A is a unitary matrix as defined earlier (see (5.7.8)).

If T is unitary, we have $TT^* = T^*T = I$, so that T is normal.

6.4.2 Example. Let V be the Euclidean plane $\mathbb{R}^2$, and let T be rotation by the angle ψ. Thus T takes the point with polar coordinates (r, θ) to the point with polar coordinates $(r, \theta + \psi)$. (You can probably convince yourself by drawing pictures of arrows that T is actually linear.) In particular,

$$T(1,0) = (\cos\psi, \sin\psi) \text{ and } T(0,1) = \left(\cos(\tfrac{\pi}{2} + \psi), \sin(\tfrac{\pi}{2} + \psi)\right)$$
$$= (-\sin\psi, \cos\psi).$$

Thus the matrix that represents T with respect to the standard basis is

$$A = \begin{bmatrix} \cos\psi & -\sin\psi \\ \sin\psi & \cos\psi \end{bmatrix}$$

(as found in Section 5.5, Problem 3). With the aid of the identity $\cos^2\psi + \sin^2\psi = 1$, we find that

$$AA^* = I$$

(hence $A^*A = I$), and we conclude that A (or equivalently T) is unitary and hence normal.

The previous statement is a bit too glib. In discussing unitary and normal operators, we have taken the underlying field to be the complex numbers, but in this example the scalars are real. In fact if we try to solve the equation $Tx = \lambda x$ for real λ and nonzero vectors x with real components, we fail completely (assuming ψ is not 0 or π). For we cannot rotate a vector x by an angle $\psi \neq 0$ or π and produce a multiple of x. Thus if we insist on using real scalars, A has no eigenvectors at all so cannot be diagonalizable, let alone orthogonally diagonalizable. We will discuss the diagonalization problem over the real field at the end of the section. But for now, let's allow complex scalars. We can then escape from our difficulties by using the matrix A to *define* a linear operator T on $\mathbb{C}^2$. If x is a vector in $\mathbb{C}^2$, then x can be expressed as $u + iv$ where u and v have real components. Then we take $Tx = Tu + iTv$ where Tu rotates u and Tv rotates v by the angle ψ. We have thus introduced complex scalars, but since A has only real entries, the conjugate transpose A^* coincides with the ordinary transpose A^t, and the above computation that $A^{-1} = A^*$ is still valid.

To produce examples of normal operators that are not unitary or self-adjoint, let A be the rotation matrix given above. Assume that ψ is not 0 or π, so that A is not Hermitian. Take $B = cA$ where c is any nonzero complex number whose magnitude is not 1. Then

$$BB^* = B^*B = |c|^2 AA^* = |c|^2 A^*A = |c|^2 I.$$

Therefore B is normal. Since $c \neq 0$, $B (= cA)$ is not Hermitian, and since $|c| \neq 1$, B is not unitary.

We now establish some basic properties of normal operators.

6.4.3 Theorem. *Let T be a normal operator on the n-dimensional inner product space V.*
(a) *$\|Tx\| = \|T^*x\|$ for every x in V; in particular T and T^* have ths same kernel.*
(b) *If c is any complex number, then $T - cI$ is normal.*
(c) *If λ is an eigenvalue of T, then the complex conjugate $\overline{\lambda}$ is an eigenvalue of T^*. Furthermore, if $Tx = \lambda x$, then $T^*x = \overline{\lambda}x$, so that if x is an eigenvector of T for the eigenvalue λ, then x is also an eigenvector of T^* for the eigenvalue $\overline{\lambda}$.*
(d) *Eigenvectors of T corresponding to distinct eigenvalues are orthogonal. Thus if $Tx = \lambda x$ and $Ty = \mu y$ where $\lambda \neq \mu$, then $\langle x, y \rangle = 0$.*
(e) *The vector x belongs to the kernel of T if and only x is orthogonal to every vector in the image of T. (We say that $\ker T$ and $\operatorname{im} T$ are orthogonal complements.)*
(f) *If $T^2 x = 0$, then $Tx = 0$.*

Proof.   (a)

$$\|Tx\|^2 = \langle Tx, Tx \rangle = \langle x, T^*Tx \rangle \quad \text{by definition of adjoint}$$

$$= \langle x, TT^*x \rangle \quad \text{since } T \text{ is normal}$$

$$= \langle x, T^{**}T^*x \rangle \quad \text{since } T^{**} = T \text{ (part (4) of (6.3.5))}$$

$$= \langle T^*x, T^*x \rangle \quad \text{by definition of adjoint}$$

$$= \|T^*x\|^2.$$

(b)   By parts (1), (2), and (5) of (6.3.5), $(T - cI)^* = T^* - \overline{c}I$. Thus

$$(T - cI)(T - cI)^* = (T - cI)(T^* - \overline{c}I) = TT^* - cT^* - \overline{c}T + |c|^2 I,$$

and

$$(T - cI)^*(T - cI) = (T^* - \overline{c}I)(T - cI) = T^*T - cT^* - \overline{c}T + |c|^2 I.$$

But $TT^* = T^*T$ since T is normal, and the result follows.

(c)   Let $S = T - \lambda I$; by (b), S is normal. If $Tx = \lambda x$, then $Sx = 0$, so $S^*x = 0$ by (a). But $S^* = T^* - \overline{\lambda}I$ as we found in the proof of (b). Thus $T^*x = \overline{\lambda}x$, as claimed.

(d) $\lambda\langle x, y \rangle = \langle \lambda x, y \rangle = \langle Tx, y \rangle = \langle x, T^*y \rangle = \langle x, \overline{\mu}y \rangle$ (by (c))
$$= \mu\langle x, y \rangle,$$
and since $\lambda \neq \mu$, we must have $\langle x, y \rangle = 0$.

(e) x is orthogonal to every vector in the image of T iff $\langle x, Ty \rangle = 0$ for all y, that is, $\langle T^*x, y \rangle = 0$ for all y. Since a vector perpendicular to everything in the entire space is perpendicular to itself and is therefore 0, the preceding condition is equivalent to $T^*x = 0$, in other words, x belongs to $\ker T^*$. But by (a), $\ker T^* = \ker T$, and the result follows.

(f) If $T^2x = 0$, then Tx belongs to $\ker T$ since $T(Tx) = T^2x = 0$, and Tx also belongs to $\operatorname{im} T$ by definition of image. But by (e), $\ker T$ and $\operatorname{im} T$ are orthogonal complements, so that $\langle Tx, Tx \rangle = 0$. Consequently, $Tx = 0$. ∎

We can now prove the fundamental theorem on normal operators.

6.4.4 Theorem. *Let T be a linear operator on the n-dimensional inner product space V over $\mathbb{C}$. Then T is unitarily diagonalizable if and only if T is normal.*

Proof. If T is unitarily diagonalizable, then with respect to some orthonormal basis, T is represented by a diagonal matrix A. But a diagonal matrix always satisfies $AA^* = A^*A$, so that T is normal.

Now assume T normal. First, we will show that T is diagonalizable by proving that the minimal polynomial of T, call it $m_T(x)$, is a product of distinct linear factors (see (6.2.5)). If $m_T(x)$ has a nonlinear factor, we can write $m_T(x) = (x-c)^2 f(x)$ for some complex number c and some monic polynomial $f(x)$. Replacing x by the operator T, we have, as discussed in (5.7.1),

$$(T - cI)^2 f(T) = m_T(T) = 0.$$

The operator $T - cI$ is normal by (6.4.3b), and we have $(T - cI)^2 f(T)v = 0$ for all $v \in V$. By (6.4.3f), $(T - cI)f(T)v = 0$ for all v, and it follows that $(T - cI)f(T) = 0$. (If a matrix B satisfies $By = 0$ for all y, then $B = 0$. To see this, take the ith entry of y to be 1, with the other components 0. Then By is column i of B, which must be 0 for all i.)

We have found a monic polynomial of lower degree than $m_T(x)$, namely $(x-c)f(x)$, that is satisfied by T, a contradiction.

Therefore T is diagonalizable, so that we have a basis consisting of eigenvectors. In fact there must be an *orthonormal* basis of eigenvectors, so that T is unitarily diagonalizable. For as in the discussion preceding (5.7.6), we can make the leap from basis to orthonormal basis if we know that eigenvectors corresponding to distinct eigenvalues are orthogonal. But this follows from (6.4.3d). ∎

Attention. In the process of checking the soundness of a proof, a mathematician always asks whether the hypotheses are actually used. Look again at the proof that if T is unitarily diagonalizable, then T is normal. Why can't we drop "unitarily"? After all, if T is diagonalizable, then T is represented with respect to some basis by a diagonalizable matrix A, and as above, we have $AA^* = A^*A$, so that T is normal. Conversely, if T is normal, then T is unitarily diagonalizable and therefore certainly diagonalizable. Thus T is diagonalizable if and only if T is normal, a false conclusion.

The solution is rather subtle. From $AA^* = A^*A$ we concluded that $TT^* = T^*T$, and in doing so we used the fact that T^* is represented by A^*. But in proving this result (see (6.3.4)), we assumed that the basis is orthonormal.

If a linear operator T is normal and therefore unitarily diagonalizable, then T can be decomposed into simpler operators, and the vector space V can be broken up into more manageable subspaces. We have the equipment available to give an informal but (I hope) clear exposition of how this process works. Suppose that the distinct eigenvalues of T are $\lambda_1, \ldots, \lambda_k$. Some of the eigenspaces may be one-dimensional. In this case we have an eigenvalue λ and an eigenvector x that spans the eigenspace of λ. Other eigenspaces may have dimension greater than 1. For example, if the eigenspace of μ is three-dimensional, then we have three orthonormal eigenvectors x, y, z spanning the eigenspace. Every vector in the eigenspace of μ will be orthogonal to all elements in all other eigenspaces. Furthermore, when T operates within a particular eigenspace, the result never leaves that space. For example, if $v = ax + by + cz$ belongs to the eigenspace of μ above, then

$$Tv = aTx + bTy + cTz = \mu(ax + by + cz) = \mu v,$$

and therefore Tv must also belong to the eigenspace. This equation also shows that within the eigenspace of μ, T simply multiplies vectors by μ.

Now let V_i be the eigenspace for the eigenvalue λ_i, $1 \le i \le k$. If $T_i x$ is the projection of x on the subspace V_i (see (5.6.9)), then

$$x \in V_i \quad \text{implies} \quad T_i x = x \tag{1}$$

(because if a vector x already belongs to a subspace, its projection on that subspace is x itself);

$$x \in V_j, \quad j \ne i \quad \text{implies} \quad T_i x = 0 \tag{2}$$

(because vectors in V_i are orthogonal to vectors in all other eigenspaces). Since we have an orthonormal basis of eigenvectors, each x has a unique representation as $x_1 + \cdots + x_k$ with the x_i orthogonal and $x_i \in V_i$ (we say that V is the *orthogonal direct sum* of the V_i). Consequently,

$$Tx = Tx_1 + \cdots + Tx_k = \lambda_1 x_1 + \cdots + \lambda_k x_k = \lambda_1 T_1 x_1 + \cdots + \lambda_k T_k x_k \quad \text{by (1).} \tag{3}$$

But $T_i x_j = 0$ for $j \ne i$ by (2), so $T_i x = \sum_{j=1}^{k} T_i x_j = T_i x_i$. Thus (3) becomes

$$Tx = \lambda_1 T_1 x + \cdots + \lambda_k T_k x, \tag{4}$$

in other words,

$$T = \lambda_1 T_1 + \cdots + \lambda_k T_k. \tag{5}$$

Furthermore, since T_j maps into V_j and T_i sends everything in V_j to 0 for $i \ne j$,

$$T_i T_j = 0, \quad i \ne j. \tag{6}$$

Finally, the representation

$$x = x_1 + \cdots + x_k = T_1 x_1 + \cdots + T_k x_k = T_1 x + \cdots + T_k x$$

shows that

$$I = T_1 + \cdots + T_k. \tag{7}$$

The analysis may be summarized as follows.

6.4.5 Spectral Theorem For Normal Operators. Let T be a normal operator on the n-dimensional complex inner product space V. Let $\lambda_1, \ldots, \lambda_k$ ($k \leq n$) be the distinct eigenvalues of T, with corresponding eigenspaces $V_1, \ldots, V_k$. If T_i is the projection onto V_i, then V is the orthogonal direct sum of the V_i and

$$I = T_1 + \cdots + T_k, \quad T = \lambda_1 T_1 + \cdots + \lambda_k T_k, \quad T_i T_j = 0, \quad i \neq j.$$

Thus from the component operators $\lambda_i T_i$ which multiply vectors in V_i by λ_i and send vectors in V_j, $j \neq i$, to 0, we can assemble the operator T.

We have seen in (6.4.2) that a matrix (or linear operator) that is unitarily diagonalizable over the complex field may fail to be orthogonally diagonalizable if we require that the underlying scalars be real. So a natural question is: Under what conditions can a matrix be orthogonally diagonalized over $\mathbb{R}$? Our analysis of the complex case answers this question as well.

6.4.6 Theorem. *Let T be a linear operator on the n-dimensional inner product space V over $\mathbb{R}$. Then T is orthogonally diagonalizable if and only if T is self-adjoint, that is, the representing matrix A is real and symmetric.*

Proof. If T is orthogonally diagonalizable, then with respect to some orthonormal basis, T is represented by a diagonal matrix A with real entries. Such a matrix must be symmetric, and therefore T is self-adjoint.

Conversely, if T is self-adjoint, then by (5.7.7), all eigenvalues of T are real. We can find eigenvectors by reducing the equation $(A - \lambda I)x = 0$ to echelon form, and this computation will involve only real numbers. Furthermore, the roots of the characteristic polynomial of A are all real, since these roots are the eigenvalues of A; see (6.2.6). Since the characteristic polynomial is a multiple of the minimal polynomial (see (6.2.6) and (6.2.7)), the roots of the minimal polynomial must be real as well. We may therefore reproduce the proof that a normal operator is unitarily diagonalizable (see (6.4.4)) to conclude that a self-adjoint operator on a real inner product space is orthogonally diagonalizable. ■

The analysis leading to the spectral theorem (6.4.5) may be reproduced verbatim for self-adjoint operators on a real inner product space.

6.4.7 Spectral Theorem For Self-Adjoint Operators on a Real Space. Let T be a self-adjoint operator on the n-dimensional real inner product space V. Let $\lambda_1, \ldots, \lambda_k$ ($k \leq n$) be the distinct eigenvalues of T, with corresponding eigenspaces $V_1, \ldots, V_k$. If T_i is the projection on V_i, then V is the orthogonal direct sum of the V_i and

$$I = T_1 + \cdots + T_k, \quad T = \lambda_1 T_1 + \cdots + \lambda_k T_k, \quad T_i T_j = 0, \quad i \neq j.$$

Problems For Section 6.4

In this problem set, all vector spaces are complex except as noted.

1. Let

$$A = \begin{bmatrix} 0 & -2 \\ 1 & 3 \end{bmatrix}.$$

Show that A is diagonalizable but not unitarily diagonalizable.

2. (Lagrange Interpolation) Suppose that a, b, c, and d are distinct complex numbers (or more generally, distinct elements of an arbitrary field). Consider the polynomial

$$f(x) = \frac{(x-b)(x-c)(x-d)}{(a-b)(a-c)(a-d)}$$

and observe that $f(a) = 1$, $f(b) = f(c) = f(d) = 0$. Similarly, if $a_0, \ldots, a_n$ are distinct elements, we can find a polynomial f_i ($0 \le i \le n$) such that $f_i(a_j)$ is 0 for $i \ne j$ and 1 for $i = j$. Now if $b_0, \ldots, b_n$ are $n + 1$ elements (not necessarily distinct), find a polynomial g of degree at most n such that $g(a_i) = b_i$ for all i.

3. Show that the operators $T_1, \ldots, T_k$ in the Spectral Theorems (6.4.5) and (6.4.7) are self-adjoint.

4. Let $T = \lambda_1 T_1 + \cdots + \lambda_k T_k$ as in (6.4.5) and (6.4.7). If f is any polynomial (with complex coefficients in (6.4.5) and real coefficients in (6.4.7)), show that $f(T) = f(\lambda_1)T_1 + \cdots + f(\lambda_k)T_k$.

Problems 5–9 give some applications of the Spectral Theorem For Normal Operators. In all problems, T is a linear operator on an n-dimensional complex inner product space.

5. Show that T is normal if and only if $T^* = g(T)$ for some polynomial g.

6. Show that T is unitary if and only if T is normal and all eigenvalues of T have magnitude 1.

7. If T is normal, show that T is self-adjoint if and only if all eigenvalues of T are real.

8. In the Spectral Theorem (either version), show that each T_i is a polynomial in T.

9. If T is normal and f is any polynomial, show that $f(T)$ is normal. If B is an orthonormal basis of eigenvectors of T corresponding to eigenvalues $\lambda_1, \ldots, \lambda_k$, show that B is also an orthonormal basis of eigenvectors of $f(T)$, corresponding to eigenvalues $f(\lambda_1), \ldots, f(\lambda_k)$.

10. Let A be the rotation matrix discussed in (6.4.2). Find the (complex) eigenvalues of A, as well as an orthonormal basis of eigenvectors. (Assume that ψ is not 0 or π.)

6.5 Existence of the Jordan Canonical Form

We will show that every n by n matrix A with complex entries is similar to a matrix in Jordan canonical form. For each eigenvalue λ of A, there will be one or more sequences of generalized eigenvectors $x_1, \ldots, x_k$ (one sequence for each Jordan block):

$$(A - \lambda I)x_1 = 0, \qquad \text{so that } x_1 \text{ is an eigenvector;}$$

$$Ax_2 = \lambda x_2 + x_1, \qquad \text{so } (A - \lambda I)^2 x_2 = 0;$$

$$\vdots$$

$$Ax_k = \lambda x_k + x_{k-1}, \qquad \text{so } (A - \lambda I)^k x_k = 0.$$

Let's call the sequence $x_1, \ldots, x_k$ a *string (of generalized eigenvectors) headed by the eigenvector x_1*.

The proof we give is due to Filippov, and the exposition follows Gilbert Strang in his book *Linear Algebra and its Applications,* Academic Press, 1980. For an alternative argument of a more computational nature, along with further discussion and references, see *Matrix Analysis* by Horn and Johnson, Cambridge, 1985.

We will argue by induction on n. When $n = 1$, there is no problem; the Jordan canonical form of the matrix $[a]$ is $[a]$ itself. Thus assume the existence of a Jordan canonical form for all m by m matrices, $m = 1, 2, \ldots, n - 1$.

We first assume that A is singular, so that the null space $N(A)$ has dimension at least 1. By the Dimension Theorem 5.4.5, the range $R(A)$ has dimension $r < n$, so by the induction hypothesis, A (more precisely the linear operator associated with A) restricted to its range has a Jordan canonical form. Thus:

(1)   There is a basis $\{w_1, \ldots, w_r\}$ for $R(A)$ such that Aw_i can be expressed either as $\lambda_i w_i$ for some eigenvalue λ_i, or as $\lambda_i w_i + w_{i-1}$.

Now suppose that the subspace $N(A) \cap R(A)$ has dimension p. Since x belongs to $N(A)$ if and only if $Ax = 0$, $N(A) \cap R(A)$ is the eigensubspace of $R(A)$ corresponding to the eigenvalue $\lambda = 0$. Thus:

(2)   In (1) there are p linearly independent eigenvectors among the w_i corresponding to the eigenvalue $\lambda = 0$, and therefore p strings headed by these eigenvectors. If w_i is the last vector in such a string, then since w_i belongs to $R(A)$ we have $w_i = Ay_i$ for some y_i. Since $Ay_i = 0y_i + w_i$, we can lengthen the string by placing y_i immediately after w_i. Note also that since $w_i \in N(A)$ we have $Aw_i = 0$.

Now the dimension of $N(A)$ is $n - r$, so in any basis of $N(A)$ there must be $n - r - p$ vectors z_i which do not belong to $N(A) \cap R(A)$. Since $z_i \in N(A)$ we have $Az_i = 0$. Thus:

(3)   There are $n - r - p$ linearly independent eigenvectors z_i belonging to $N(A)$ but not to $R(A)$. (These vectors will produce Jordan blocks of order 1.)

We form a Jordan basis for the entire space by collecting the r vectors w_i in step 1, the p vectors y_i in step 2, and the $n - r - p$ vectors z_i in step 3. The Jordan condition $Ax_i = \lambda x_i$ or $\lambda x_i + x_{i-1}$ is satisfied (since it holds in each of the 3 steps), so we need only show that this collection of n vectors is linearly independent. Assume then that there are scalars a_i, b_i, c_i such that

$$\sum a_i w_i + \sum b_i y_i + \sum c_i z_i = 0. \tag{4}$$

Apply A to both sides to get

$$\sum a_i(\lambda_i w_i \text{ or } \lambda_i w_i + w_{i-1}) + \sum b_i A y_i = 0, \tag{5}$$

where terms that turn out to be 0 are omitted.

In (5), each $A y_i$ is one of the w_i in (4). But $A y_i$ will never coincide with any of the w_i appearing in the first summation of (5), because if $w_i = A y_i$ then $A w_i = 0$, as we observed in (2). Furthermore, $A y_i$ is not one of the w_{i-1}, since w_{i-1} is not the last vector in a string. By linear independence of the w_i, all the b_i must be 0.

Suppose that $\{v_1, \ldots, v_p\}$ is a basis for $N(A) \cap R(A)$, and we extend this set so that $\{v_1, \ldots, v_p, z_1, \ldots, z_{n-r-p}\}$ is a basis for $N(A)$ (see (5.3.3c)). By the Gram-Schmidt Process (5.6.8), we can find an orthonormal basis $\{v_1', \ldots, v_p'\}$ for $N(A) \cap R(A)$, extended to an orthonormal basis $\{v_1', \ldots, v_p', z_1', \ldots, z_{n-r-p}'\}$ for $N(A)$.

In particular, each z_i' is orthogonal to the subspace $N(A) \cap R(A)$. To avoid extra notation, we may as well assume that each of the original z_i's is orthogonal to $N(A) \cap R(A)$.

Now by (4), we have $z = \sum c_i z_i = -\sum a_i w_i$. Since z is a linear combination of the z_i, and of the w_i as well, we have $z \in N(A) \cap R(A)$. But each z_i is orthogonal to $N(A) \cap R(A)$, and therefore so is z. We conclude that z is perpendicular to itself, so that $z = 0$. This forces all c_i to be 0, by linear independence of the z_i. Finally, the linear independence of the w_i yields $a_i = 0$ for all i, completing the proof under the assumption that A is singular.

If A is nonsingular, let λ be any eigenvalue of A, and let $B = A - \lambda I$, which is singular (since $\det B = 0$). We have proved that B is similar to a matrix J_0 in Jordan canonical form, so that there is an invertible matrix P such that $P^{-1} B P = J_0$. But then

$$P^{-1} A P = P^{-1}(B + \lambda I)P = P^{-1} B P + \lambda I = J_0 + \lambda I.$$

But $J = J_0 + \lambda I$ is formed from J_0 by adding λ to every main diagonal entry of J_0. Thus J is still in Jordan canonical form (the eigenvalues of B have been increased by λ to produce the eigenvalues of A), and the same basis of generalized eigenvectors that worked for B will work for A, because we are using the same basis-changing matrix P.∎

Problems For Section 6.5

1. Where in the proof did we use the assumption that the scalars were complex?

(The Gram-Schmidt Process is one candidate, but everything in Section 5.6 can be done with real scalars; whenever a complex conjugate $\bar{b}$ appears, replace it by b. However, there is one place in the proof where the complex numbers are really needed.)

2. This problem is not about the proof of existence of the Jordan canonical form. Instead, I would like to end the book with a problem that brings together many of the ideas in Chapters 5 and 6.

Let T and S be linear operators on the n-dimensional vector space V over the complex numbers. Assume that T and S are diagonalizable, and that T and S commute, in other words, $TS = ST$.

(a) If x is an eigenvector of S, show that Tx is also an eigenvector of S for the same eigenvalue. That is, if $Sx = \lambda x$, then $S(Tx) = \lambda(Tx)$.

(b) If W is any eigenspace of S, show that W is T-invariant, that is, if x belongs to W, so does Tx. Thus if we restrict T to W, we get a legal linear operator on W.

(c) Show that the restriction T_W of T to W is diagonalizable, and if T is unitarily diagonalizable, so is T_W.

(d) Show that T and S are simultaneously diagonalizable, that is, there is a basis $\{x_1, \ldots, x_n\}$ where the x_i are eigenvectors of both operators T and S.

(e) If S and T are unitarily diagonalizable and $TS = ST$, show that T and S are simultaneously unitarily diagonalizable; that is, there is an orthonormal basis $\{x_1, \ldots, x_n\}$ where the x_i are eigenvectors of both T and S.

(f) Show that if T and S are simultaneously diagonalizable linear operators on V, they must commute.

Appendix

An Application Of Linear Algebra

Virtually every branch of science uses linear algebra. Here is an application that is of interest in many fields. A *finite Markov chain* is a system with *states* $s_1, \ldots, s_r$ and *transition probabilities* p_{ij}, $i, j = 1, \ldots, r$. Starting in an initial state at time $t = 0$, the system moves from one state to another at subsequent times $t = 1, 2, \ldots$. If the system is in state i at a given time, the probability that it will move to state j at the next transition is p_{ij}. (We allow $j = i$, so that p_{ii} can be greater than zero.)

The matrix A with entries p_{ij} is called the *transition matrix* of the chain. It is an example of a *stochastic matrix*: the entries are nonnegative and the sum across each row is 1.

If we start in state i at $t = 0$, what is the probability $p_{ij}^{(2)}$ that we will be in state j after *two* transitions? One way this can happen is to move to state k at $t = 1$ and then move from state k to state j at time $t = 2$. The probability that this will occur is $p_{ik}p_{kj}$. But k can be any integer from 1 to r, and we must add all of the corresponding probabilities. The result is

$$p_{ij}^{(2)} = \sum_{k=1}^{r} p_{ik}p_{kj},$$

which is the ij entry of the matrix A^2.

Thus the entries of A^2 are the two-step transition probabilities. Similarly, we can consider three-step transition probabilities $p_{ij}^{(3)}$. If we start in s_i at $t = 0$, one way of arriving at s_j at $t = 3$ is to be in s_k at $t = 2$ and move from s_k to s_j at $t = 3$. This event has probability $p_{ik}^{(2)}p_{kj}$, and consequently

$$p_{ij}^{(3)} = \sum_{k=1}^{r} p_{ik}^{(2)}p_{kj},$$

the ij entry of $A^2 A = A^3$.

An induction argument shows that if $p_{ij}^{(n)}$ is the probability, starting in s_i, of being in s_j n steps later, then $p_{ij}^{(n)}$ is the ij entry of A^n. Thus to compute n-step transition probabilities, we must calculate the nth power of the transition matrix A. This is quite a tedious chore for large n. But if A is diagonalizable (in particular, if A has distinct eigenvalues), and the eigenvalues and eigenvectors of A are found, then all powers of A can be computed efficiently, as follows.

Let P be a nonsingular matrix such that $P^{-1}AP = D$, a diagonal matrix whose main diagonal entries are the eigenvalues λ_i ($i = 1, \ldots, r$) of A. Then $A = PDP^{-1}$, and if we begin to compute the powers of A, a pattern emerges quickly:

$$A^2 = AA = PDP^{-1}PDP^{-1} = PD^2P^{-1},$$

$$A^3 = A^2A = PD^2P^{-1}PDP^{-1} = PD^3P^{-1},$$

and by induction,

$$A^n = PD^nP^{-1}.$$

But since D is diagonal, so is D^n, and the main diagonal entries of D^n are λ_i^n, $i = 1, \ldots, r$. Once the eigenvalues and eigenvectors have been found, the matrix P can be taken to have eigenvectors as columns. The computation of A^n has been reduced to finding the nth powers of the λ_i, followed by a matrix inversion and two matrix multiplications, one of which is easy (because D^n is diagonal).

Solutions to Problems

Section 1.1

1.

A	B	$A \vee B$	$\neg(A \vee B)$	$\neg A$	$\neg B$	$(\neg A) \wedge (\neg B)$
T	T	T	F	F	F	F
T	F	T	F	F	T	F
F	T	T	F	T	F	F
F	F	F	T	T	T	T

2. To prove the first law, note that the left side is true iff $A_1 \wedge \cdots \wedge A_n$ is false, which happens iff at least one A_i is false, i.e., at least one $(\neg A_i)$ is true, equivalently, the right side is true. For the second law, note that the left side is true iff $A_1 \vee \cdots \vee A_n$ is false, which happens iff all A_i are false, i.e., all $(\neg A_i)$ are true, equivalently, the right side is true.

3.

A	B	$A \Rightarrow B$	$\neg A$	$(\neg A) \vee B$
T	T	T	F	T
T	F	F	F	F
F	T	T	T	T
F	F	T	T	T

4.

A	$\neg A$	$A \vee (\neg A)$	$A \wedge (\neg A)$
T	F	T	F
F	T	T	F

5. The left side is true iff A and (either B or C) are true. The right side is true iff either (A and B) or (A and C) is true, in other words, A is true in conjunction with either B or C. Thus the two sides have the same truth table. (If you are not comfortable with this reasoning, construct the complete truth tables for $A \wedge (B \vee C)$ and $(A \wedge B) \vee (A \wedge C)$, and verify that they are identical.)

6. The left side is true iff either A or (B and C) is true. The right side is true iff both (A or B) and (A or C) are true. This will happen if A is true, but if A is false, both B and C must be true (a proof by cases; see Section 1.3). Thus the right side is true iff either A or (B and C) is true. As in Problem 5, this can be verified by a truth table.

7. Going from Problem 5 to Problem 6 gives a concrete example with the essential features of the general case, so let's do it this way rather than use messy formal notation. Having established the result of Problem 5, take the negation of both sides, using the DeMorgan Laws. We get

$$\neg\big[A \wedge (B \vee C)\big] \Leftrightarrow \neg\big[(A \wedge B) \vee (A \wedge C)\big]$$

$$\big[(\neg A) \vee \neg(B \vee C)\big] \Leftrightarrow \Big(\big[\neg(A \wedge B)\big] \wedge \big[\neg(A \wedge C)\big]\Big)$$

$$\Big[(\neg A) \vee \big((\neg B) \wedge (\neg C)\big)\Big] \Leftrightarrow \Big(\big[(\neg A) \vee (\neg B)\big] \wedge \big[(\neg A) \vee (\neg C)\big]\Big).$$

This is the result of Problem 6, except that each proposition A, B, C is replaced by its negation. But A, B, and C are *arbitrary* propositions, which is a key point; as A ranges over all possible propositions, so does $\neg A$. (A similar but perhaps more familiar statement is that if x ranges over all real numbers, so does $-x$; if you want $-x$ to equal y, take $x = -y$). Thus the result of Problem 6 holds in general. Notice also that if a tautology T appears in the original statement, taking the negation changes it to F, and similarly a contradiction F is changed to T.

Section 1.2

1. $\forall x \, \exists N \, (N > x)$

2. $\exists x \, \forall N \, (N \leq x)$, which says that there is a real number x that is at least as big as every integer (false!).

Section 1.3

1. True for $n = 1$, since $1(2)/2 = 1$. If true for n, then

$$1 + 2 + \cdots + n = n(n+1)/2 \qquad \text{by the induction hypothesis}$$

$$n + 1 = n + 1 \qquad \text{(an identity), so}$$

$$1 + 2 + \cdots + n + 1) = \big[n(n+1)/2\big] + (n+1) = (n+1)\big[(n/2) + 1\big]$$

$$= (n+1)(n+2)/2.$$

Thus the statement is true for $n+1$, and therefore the result holds for all n, by mathematical induction.

2. True for $n = 1$, since $2^{2(1)} - 1 = 4 - 1 = 3$. If $2^{2n} - 1$ is divisible by 3, consider

$$2^{2(n+1)} - 1 = 2^{(2n+2)} - 1 = (4)2^{2n} - 1 = (3)2^{2n} + (2^{2n} - 1).$$

By the induction hypothesis, $2^{2n} - 1$ is divisible by 3, and it follows that $2^{2(n+1)} - 1$ is the sum of two numbers divisible by 3, and consequently is divisible by 3. The induction argument is therefore complete.

3. True for $n = 1$, since $11^1 - 4^1 = 7$. If $11^n - 4^n$ is divisible by 7, then

$$11^{n+1} - 4^{n+1} = 11(11^n) - 4(4^n) = 11(11^n - 4^n) + (11 - 4)4^n,$$

which (using the induction hypothesis) is the sum of two numbers divisible by 7. The result follows.

4. True for $n = 1$, since $1^2 = 1(2)(3)/6$. If true for n, then by the induction hypothesis,

$$
\begin{aligned}
1^2 + 2^2 + \cdots + n^2 + (n+1)^2 &= \frac{n(n+1)(2n+1)}{6} + (n+1)^2 \\
&= (n+1)\left(\frac{2n^2 + n}{6} + n + 1\right) \qquad \text{by factoring} \\
&= \frac{(n+1)(2n^2 + 7n + 6)}{6} \qquad \text{by algebra} \\
&= \frac{(n+1)(n+2)(2n+3)}{6} \qquad \text{by more algebra.}
\end{aligned}
$$

Since $2n + 3 = 2(n+1) + 1$, the induction step is proved.

5. The assertion is true for a postage of $n = 35$ cents, since we can pay with seven 5-cent stamps. If the result holds for a postage of n cents ($n \geq 35$), consider a postage of $n + 1$. In case 1, a postage of n can be paid with all 5's, and it takes at least seven of them since $n \geq 35$. If we replace seven 5's by four 9's, we have paid for $n + 1$ using only 5's and 9's. In case 2, postage n is paid using at least one 9. To pay for n+1 in this case, replace the 9 by two 5's, and again we have paid for $n + 1$ using only 5's and 9's. This completes the induction step.

Section 1.4

1. We have $x \in (\bigcap_i A_i)^c$ iff $x \notin \bigcap_i A_i$ iff it is not the case that x belongs to A_i for all i iff for at least one i, $x \notin A_i$ iff $x \in \bigcup_i (A_i^c)$.

2. We have $x \in A \cup (\bigcap_i B_i)$ iff $x \in A$ or $x \in B_i$ for all i iff for all i, $x \in A$ or $x \in B_i$ iff for all i, $(x \in A$ or $x \in B_i)$ iff $x \in \bigcap_i (A \cup B_i)$.

3. We must show that A has no members. But if $x \in A$ then by hypothesis, x belongs to the empty set, which is impossible.

4. If $i \neq j$, then $B_i \cap B_j \subseteq A_i \cap A_j = \varphi$. By Problem 3, $B_i \cap B_j = \varphi$.

5. No. For example, let $A = \{1, 2, 3, 4\}$, $B = \{1, 2\}$, $C = \{1, 4\}$. Then $A \cup B = A \cup C = A$, but $B \neq C$.

6. No. For example, let $A = \{1, 2, 3, 4\}$, $B = \{1, 2, 5\}$. Then $A \cup (B \backslash A) = \{1, 2, 3, 4\} \cup \{5\} \neq B$.

7. $A \cup (B \setminus A) = A \cup B$. For if $x \in A \cup (B \setminus A)$, then x belongs to A or x belongs to B but not A, so that $x \in A \cup B$. Conversely, if $x \in A \cup B$, then it is convenient to do a proof by cases:

Case 1. $x \in A$; then certainly $x \in A \cup (B \setminus A)$.
Case 2. $x \notin A$; then, since $x \in A \cup B$, we must have $x \in B$, so that $x \in B \setminus A$.

(A Venn diagram may be useful in visualizing the result.)

8. The Distributive Law provides a concrete example with all the features of the general case. In the original Distributive Law $A \cap (\bigcup_i B_i) = \bigcup_i (A \cap B_i)$, take the complement of both sides and use the DeMorgan Laws to obtain $A^c \cup (\bigcap_i B_i^c) = \bigcap_i (A^c \cup B_i^c)$. Since the sets A and B_i are arbitrary, we may replace A^c by A and B_i^c by B_i to obtain the second Distributive Law of Problem 2. Notice that if Ω appears in the original identity, taking the complement changes Ω to $\varnothing$. Similarly, $\varnothing$ is replaced by Ω.

9. $A \subseteq B$ iff $(x \in A \Rightarrow x \in B)$ iff $((x \in A) \Leftrightarrow (x \in A \text{ and } x \in B))$ iff $((x \in B) \Leftrightarrow (x \in A \text{ or } x \in B))$, and the result follows.

Section 1.5

1. If $x^3 = y^3$, we may take cube roots to conclude that $x = y$, so f is injective. Any real number y has a real cube root $x = y^{1/3}$, so f is surjective.

2. f is neither injective nor surjective, by an analysis similiar to that in the text for $f(x) = x^2$.

3. $h(x) = g(f(x))$, where $f(x) = x^2 + 1$ and $g(y) = y^{10}$.

4. If A consists of a single point then f is injective, and if B consists of a single point (necessarily c), then f is surjective. These are the only circumstances.

5. If $A = \{a_1, \ldots, a_m\}$, then B has at least m distinct points $f(a_1), \ldots, f(a_m)$, so $m \leq n$.

6. If $B = \{b_1, \ldots, b_n\}$ then for each i there is a point $a_i \in A$ such that $f(a_i) = b_i$. The elements a_i are distinct, for otherwise the function f would map the same point to two different images in B, which is impossible. Thus A has at least n distinct points, so that $m \geq n$.

7. In view of (1.5.5(a)), we need only prove that $f^{-1}[f(C)]$ is a subset of C. If $x \in f^{-1}[f(C)]$, then $f(x) \in f(C)$, so that $f(x) = f(y)$ for some $y \in C$. Since f is injective we have $x = y$, and therefore $x \in C$.

8. In view of (1.5.5(b)), we need only prove that D is a subset of $f[f^{-1}(D)]$. If $y \in D$, then since f is surjective we have $y = f(x)$ for some $x \in A$. But then $f(x) = y \in D$, so $y = f(x)$ with $x \in f^{-1}(D)$; that is, $y \in f[f^{-1}(D)]$.

9. In view of (1.5.5(d)), we need only prove that the intersection of the $f(A_i)$ is a subset of $f(\bigcap_i A_i)$. If $y \in \bigcap_i f(A_i)$, then for each i we have $y = f(x_i)$ for some $x_i \in A_i$. Since f is injective, all the x_i are equal (to x, say); hence $y = f(x)$ with $x \in \bigcap_i A_i$, and the result follows.

Section 1.6

1. R is reflexive (W and W certainly begin with the same letter), symmetric (if W and V begin with the same letter, so do V and W) and transitive (if W and V begin with the same letter, and V and U begin with the same letter, then W and U begin with the same letter). If W begins with a, the equivalence class of W consists of all words beginning with a. Thus there are 26 equivalence classes, one for each possible letter.

2. If aRb, then bRa by symmetry, so $a = b$ by antisymmetry. Conversely, if $a = b$, then aRb by reflexivity. Thus aRb if and only if $a = b$.

3. The argument of Problem 2 uses reflexivity, which is no longer assumed.

4. Let $A = \{1, 2, 3\}$ and let R consist of the ordered pairs $(1, 1)$ and $(2, 2)$. Then R is symmetric and antisymmetric, but $(3, 3) \notin R$, so that R is not equality.

5. If R is relation that is reflexive, symmetric and antisymmetric, then R is the equality relation. The argument of Problem 2 goes through in this case.

6. No. If a and b are maximal and R is total, then aRb or bRa. If, say, aRb, then since a is maximal we have $a = b$.

7. The inclusion relation is reflexive ($A \subseteq A$), antisymmetric (if $A \subseteq B$ and $B \subseteq A$ then $A = B$), and transitive (if $A \subseteq B$ and $B \subseteq C$ then $A \subseteq C$). The relation is not total (unless W has at most one element). For example, if $A = \{1, 2, 3\}$ and $B = \{2, 3, 4\}$ then A is not a subset of B and B is not a subset of A.

8. (a) If $x \in A_j$, then certainly $x \in A_i$ for at least one i, so $A_j \subseteq B$.

(b) We must show that if each $A_i \subseteq C$, then $\bigcup_i A_i \subseteq C$. But this follows directly from the definition of union.

9. (a) If $x \in B$, then x belongs to every A_i, so $B \subseteq A_i$ for all i.

(b) We must show that if $C \subseteq A_i$ for every i, then $C \subseteq \bigcap_i A_i$. But this follows directly from the definition of intersection.

Section 2.1

1. A bijective function from A to A corresponds to a permutation of A, and by (2.1.2), the total number of permutations is $n!$.

2. We have n choices for $f(a)$, where a ranges over the k elements of A. The total number of functions is $(n)(n)\cdots(n) = n^k$.

3. Once an element $f(a) \in B$ is chosen, it cannot be used again. Therefore the number of injective functions is

$$(n)(n-1)\cdots(n-k+1) = \frac{n!}{(n-k)!}.$$

4. By Problem 2, the number of functions from A to $\{0,1\}$ is 2^n.

5. Suppose that 1 and 4 go to R_1, 5 to R_2, and 2 and 3 to R_3. This corresponds to the sequence $R_1 R_3 R_3 R_1 R_2$. In general, we are counting generalized permutations of R_1, R_2, and R_3 in which R_1 occurs twice, R_2 once and R_3 twice. The result is $\frac{5!}{2!1!2!} = 30$.

6. By the formula for generalized permutations, the number of assignments is

$$\frac{n!}{k_1! \cdots k_r!}.$$

7. The assignment of Problem 5 yields the partition $\{1,4\}$, $\{5\}$, $\{2,3\}$. But the assignment in which 1 and 4 go to R_3, 5 to R_2, and 2 and 3 to R_1, yields the same partition, since we get the same collection of disjoint subsets whose union is $\{1,2,3,4,5\}$. Because there are two rooms of the same size, the computation of Problem 5 overcounts by a factor of 2, and the correct answer is $30/2 = 15$.

8. Suppose we have two subsets S_1 and S_2 of size 5, four subsets T_1, T_2, T_3, and T_4 of size 3, and one subset U_1 of size 2. This can be converted into a room assignment by permuting S_1 and S_2, and then permuting T_1, T_2, T_3, and T_4. (There is only one permutation of the single symbol U_1.) For example, $S_2 S_1 T_3 T_4 T_2 T_1$ corresponds to sending the people in S_2 to room R_1, the people in S_1 to R_2, the people in T_3 to R_3, T_4 to R_4, T_2 to R_5, and T_1 to R_6. Thus the number of partitions times $2!4!1!$ is the number of room assignments, so the correction factor is $2!4!1!$.

9. By the same reasoning as in Problem 8, we obtain

$$\frac{n!}{k_1! \cdots k_r! t_1! \cdots t_m!}.$$

10. We are counting the number of nonnegative integer solutions of $x_1 + x_2 + x_3 + x_4 + x_5 = 10$, which is

$$\binom{10 + 5 - 1}{10}.$$

Section 2.2

1. $(1 + 1)^n = \sum_{k=0}^{n} \binom{n}{k} 1^k 1^{n-k}$, and the result follows.

2. $(-1 + 1)^n = \sum_{k=0}^{n} \binom{n}{k} (-1)^k 1^{n-k}$, and the result follows.

3. By Problem 4 of Section 2.1, there are 2^n subsets of a set A with n elements. But by (2.1.4), there are $\binom{n}{k}$ k-element subsets of A. Sum from $k = 0$ to n to obtain the desired identity.

4. The desired identity is

$$\frac{n!}{k!(n-k)!} = \frac{(n-1)!}{(k-1)!(n-k)!} + \frac{(n-1)!}{k!(n-k-1)!}.$$

Multiply by $k!(n-k)!$ to obtain $n! = k(n-1)! + (n-k)(n-1)! = n(n-1)!$, which is valid. The steps of this argument may be reversed to establish the original identity.

5. There are $\binom{n}{k}$ k-element subsets of $\{1, 2, \ldots, n\}$. Consider any fixed element of $\{1, 2, \ldots, n\}$, say n. If S is a k-element subset, there are two possibilities:

Case 1. $n \in S$. Then there are $k - 1$ other elements of S, to be chosen from the integers $1, 2, \ldots, n - 1$. The number of such subsets is $\binom{n-1}{k-1}$.

Case 2. $n \notin S$. Then S is a k-element subset of $\{1, 2, \ldots, n - 1\}$, and the number of such subsets is $\binom{n-1}{k}$.

Now any k-element subset falls into one of the two cases (but not both), and therefore the total number of k-element subsets is the sum of the number of subsets in case 1 plus the number in case 2. The result follows.

6. The sum of all the coefficients in the multinomial expansion of $(a_1 + \cdots + a_r)^n$ may be obtained by setting all $a_i = 1$ (cf. Problem 1). The sum of the coefficients is therefore r^n. When $r = 3$ and $n = 4$, we get $3^4 = 81$, as expected.

Section 2.3

1. We must place i in position i, and the remaining $n - 1$ integers $1, 2, \ldots, i - 1, i + 1, \ldots, n$ can be permuted arbitrarily. Thus $N(A_i)$ is the number of permutations of a set with $n - 1$ members, which is $(n - 1)!$

2. We must place $i_1, \ldots, i_k$ in their natural positions, and we can then permute the remaining $n - k$ integers arbitrarily. There are $(n - k)!$ ways of doing this.

3. The number $d(n)$ of derangements is the total number of permutations minus the number of permutations in which at least one integer stands in its natural position. Thus $d(n) = n! - N(A_1 \cup \cdots \cup A_n)$, and we compute $N(A_1 \cup \cdots \cup A_n)$ with the aid of PIE_n. There are $\binom{n}{i}$ terms involving intersections of i of the sets A_j. Terms involving an even number of intersections appear with a minus sign, and by Problem 2, each term is $(n - i)!$ in absolute value. Therefore

$$d(n) = n! - \sum_{i=1}^{n} (-1)^{i-1} \binom{n}{i} (n - i)! = \sum_{i=0}^{n} (-1)^i \binom{n}{i} (n - i)!.$$

The alternative expression for $d(n)$ follows from the identity

$$\binom{n}{i} = \frac{n!}{i!(n - i)!}.$$

4. By Problem 3,

$$\left| d(n) - \frac{n!}{e} \right| = n! \left| \sum_{i=n+1}^{\infty} \frac{(-1)^i}{i!} \right|.$$

Now an alternating series whose terms decrease in magnitude must be less than the first term in absolute value, so

$$\left| d(n) - \frac{n!}{e} \right| < \frac{n!}{(n + 1)!} = \frac{1}{n + 1} \leq \frac{1}{2},$$

and the result follows.

5. $N(A_i)$ is the number of functions from a set with k elements to a set with $n - 1$ elements (one of the original n elements, namely i, is excluded). By Problem 2 of Section 2.1, $N(A_i) = (n - 1)^k$.

6. We are counting the number of functions from a set with k elements to a set with $n - r$ elements (r of the original n elements are excluded). The result is $(n - r)^k$.

7. The number $S(k, n)$ of surjective functions is the total number of functions minus the number of functions f such that some integer $i \in \{1, \ldots, n\}$ is missing from the image of f. Thus $S(k, n) = n^k - N(A_1 \cup \cdots \cup A_n)$, and we compute $N(A_1 \cup \cdots \cup A_n)$ with the aid of PIE_n. There are $\binom{n}{i}$ terms involving intersections of i of the sets. Terms involving an even number of intersections appear with a minus sign, and by Problem 6, each term is $(n-i)^k$ in absolute value. Therefore

$$S(k,n) = n^k - \sum_{i=1}^{n} (-1)^{i-1} \binom{n}{i} (n-i)^k$$

$$= \sum_{i=0}^{n} (-1)^i \binom{n}{i} (n-i)^k.$$

8. A partition of $\{1,\ldots,8\}$ into four disjoint nonempty subsets gives rise to $4! = 24$ surjective functions; there are 4 possible choices for $f(1) (= f(2))$, and then 3 possible choices for $f(3) (= f(4) = f(5))$, and so on. For example, we might choose $f(1) = f(2) = 3$, $f(3) = f(4) = f(5) = 1$, $f(6) = 4$, $f(7) = f(8) = 2$. The correct statement is that the number of surjective functions from $\{1,\ldots,8\}$ to $\{1,2,3,4\}$ is $4!$ times the number of partitions of $\{1,\ldots,8\}$ into four disjoint nonempty subsets.

9. $S(k,n) = n!P(k,n)$. The reasoning is the same as in the concrete example of Problem 8.

10. $S(k,n) = 3^4 - \binom{3}{1}2^4 + \binom{3}{2}1^4 - \binom{3}{3}0^4 = 81 - 48 + 3 - 0 = 36$

$$P(k,n) = \frac{S(k,n)}{n!} = \frac{36}{3!} = 6.$$

The partitions are

$$\{1,2\},\{3\},\{4\}$$
$$\{1,3\},\{2\},\{4\}$$
$$\{1,4\},\{2\},\{3\}$$
$$\{2,3\},\{1\},\{4\}$$
$$\{2,4\},\{1\},\{3\}$$
$$\{3,4\},\{1\},\{2\}.$$

Section 2.4

1. There is no way to guarantee that the number r selected is rational.

2. We give a proof by mathematical induction. The $n = 2$ case follows from the diagonal scheme that we used to count the rationals. If $A_1 = \{a_1, a_2, \ldots\}$ and $A_2 = \{b_1, b_2, \ldots\}$, we simply replace the rational number i/j by the ordered pair (a_i, b_j). If we have proved that the Cartesian product of $n-1$ countable sets is countable, then the result for n sets follows because an ordered n-tuple $(x_1, x_2, \ldots, x_n)$ can be regarded as an ordered pair $((x_1, \ldots, x_{n-1}), x_n)$. The result then follows from the induction hypothesis and the $n = 2$ case.

3. Let

$$x = \frac{r_1 + r_2}{2};$$

then $r_1 < x < r_2$, so x must occur after r_1 but before r_2 on the list. This is a contradiction, since we are given that r_1 is followed immediately by r_2. Alternatively, simply observe that there is no smallest positive rational, so the list cannot even get started.

4. Let a_1 be any element of A, and set $f(1) = a_1$. Since A is infinite, it must contain an element $a_2 \neq a_1$; set $f(2) = a_2$. Since A is infinite, it must contain an element a_3 with $a_3 \neq a_1$ and $a_3 \neq a_2$; set $f(3) = a_3$. We continue in this fashion, performing an inductive procedure (compare the proof of (1.6.5)). At step n we have distinct points $a_1, \ldots, a_n$, with $f(i) = a_i$, $1 \leq i \leq n$. If we define $f : Z^+ \to A$ by $f(n) = a_n$, $n = 1, 2, \ldots$, then f is a one-to-one mapping of Z^+ into A.

Section 3.1

1. By (i), d divides both a and b, so by (ii), d divides e. A symmetrical argument shows that e divides d. Thus $|d| \leq |e|$ and $|e| \leq |d|$, so $|d| = |e|$.

2. If e is any positive integer that divides both a and b, then e divides d by definition of d, so $e \leq |d|$, and the result follows.

3.

i	q_i	s_i	t_i	r_i
-1		1	0	770
0		0	1	84
1	9	1	-9	14

$\gcd(770, 84) = 14$, and $1(770) - 9(84) = 14$.

4.

i	q_i	s_i	t_i	r_i
-1		1	0	232
0		0	1	14
1	16	1	-16	8
2	1	-1	17	6
3	1	2	-33	2

$\gcd(232, 14) = 2$, and $2(232) - 33(14) = 464 - 462 = 2$.

5. Not unique. If $sa + tb = d$, then $(s + kb)a + (t - ka)b = sa + tb = d$, so there are infinitely many solutions.

Section 3.2

1. $10561485 = (3)(5)(11^3)(23)^2$

2. N can be written as a product of primes, in particular, N has at least one prime factor p, which must be one of the p_i. But then p divides N and p divides $p_1 p_2 \ldots p_k$; hence p divides 1, a contradiction.

3. If $N = t(n+1)! + 1$, then $N + r - 1 = t(n+1)! + r$, which is divisible by r for $r = 2, 3, \ldots, n+1$, which implies that $N + r - 1$ is composite. Thus $N + 1, \ldots, N + n$ are all composite.

4. If c is any composite number between 1 and n, then c must have a prime factor $p \leq \sqrt{n}$ (otherwise $c = ab$ where both a and b exceed $\sqrt{n}$, so $c > n$, a contradiction). Thus c will be removed from the list.

5. If p^e appears in the prime factorization of a, then by the Unique Factorization Theorem, p^{ke} must appear in the prime factorization of a^k. Thus all exponents in the prime factorization of a^k (and similarly b^k) are multiples of k, and therefore all exponents in the prime factorization of n are multiples of k. It follows that $\sqrt[k]{n}$ is an integer, contradicting the hypothesis.

6. (a) The least common multiple is $m = p_1^{g_1} \ldots p_k^{g_k}$ where $g_i = \max(e_i, f_i)$. The argument is exactly the same as in Theorem 3.2.6, with all inequalities reversed and divisors replaced by multiples.

(b) In view of part (a) and (3.2.6), we must show that

$$p^e p^f = p^{\min(e,f)} p^{\max(e,f)}$$

or equivalently, $e + f = \min(e, f) + \max(e, f)$. But this is always true (the sum of two numbers is the smaller plus the larger).

7. If t is any positive integer that is a multiple of both a and b, then by definition of m, we have $m | t$, so $|m| \leq t$, and the result follows.

8. $\gcd(a, b) = (2^2)(5)(13)$, $\text{lcm}(a, b) = (2^3)(5^2)(7)(13^2)$.

Section 3.3

1. $3(0) = 0$, $3(1) = 3$, $3(2) = 1$ (note $6 \equiv 1 \bmod 5$), $3(3) = 4$, $3(4) = 2$. Since $3(2) = 1$ in $\mathbb{Z}_5$, the multiplicative inverse of 3 is 2.

2. $1^{-1} = 1$, $2^{-1} = 3$, $3^{-1} = 2$, $4^{-1} = 4$.

3. $F[X]$ is a commutative ring because polynomials can be added, subtracted and multiplied and the result will still be a polynomial. (Formally, axioms (A1)–(A5) and (M1)–(M5) must be checked.) In fact $F[X]$ is an integral domain. To see this, suppose $f(X)g(X) = (a_n X^n + \cdots + a_0)(b_m X^m + \cdots + b_0) = 0$. If neither $f(X)$ nor $g(X)$ is 0, then we have nonzero leading coefficients a_n and b_m whose product is 0, contradicting the fact that F is a field. $F[X]$ is not a field because $f(X)/g(X)$ is in general not a polynomial (for example, let $f(X) = X + 2$ and $g(X) = X + 1$).

4. We obtain the field $F(X)$ of *rational functions* $f(X)/g(X)$, where $f(X)$ and $g(X)$ are polynomials with coefficients in F, and $g(X) \neq 0$. Since the sum, difference, product or quotient (with nonzero denominator) of rational functions is also a rational function, $F(X)$ is a field.

Section 3.4

1. As in (3.4.2), we find that $4(37) - 7(21) = 1$, and it follows that -7 is a multiplicative inverse of 21 mod 37. We are free to replace -7 by the canonical representative $-7 + 37 = 30$.

2. As in (3.4.2), we find that $3(127) - 38(10) = 1$, so -38 is a multiplicative inverse of 10 mod 127, and we can replace -38 by $-38 + 127 = 89$. If $10x \equiv 7 \bmod 127$, then $x \equiv (10)^{-1}(7) \equiv 89(7) \equiv 115 \equiv -12 \bmod 127$.

3. We have $1 \equiv 1 \bmod 9$, $10 \equiv 1 \bmod 9$, $10^2 = 10(10) \equiv 1(1) = 1 \bmod 9, \ldots$, $10^{n-1} \equiv 1 \bmod 9$, so $N \equiv a_1 + a_2 + \cdots + a_n \bmod 9$.

4. We have $1 \equiv 1 \bmod 11$, $10 \equiv -1 \bmod 11$, $10^2 \equiv (-1)^2 = 1 \bmod 11$, $10^3 \equiv (-1)^3 = -1 \bmod 11, \ldots, 10^{n-1} \equiv (-1)^{n-1} \bmod 11$. Thus $N \equiv a_1 - a_2 + a_3 - a_4 + \cdots \bmod 11$.

5. We have

$$N = (a_1 + a_2 10^1 + \cdots + a_r 10^{r-1}) + (a_{r+1} 10^r + \cdots + a_n 10^{n-1})$$
$$= A + B,$$

and since $M = 2^r$ and 2 divides 10, M divides B. Thus M divides N if and only if M divides A, as asserted.

6. Let p be a prime factor of N. Then p cannot be any of the p_i, for if p_i were to divide N, the equation $N = 4p_1 \cdots p_k - 1$ implies that p_i divides 1, which is impossible. Since $p_1, \ldots, p_k$ constitute all the primes $\equiv 3 \bmod 4$, p must be congruent to 1 mod 4. (If $p \equiv 0 \bmod 4$, then 4 divides p, which is impossible because p is prime. If $p \equiv 2 \bmod 4$, then p is even, so that $p = 2$. This cannot happen because N is an odd number.) Since the product of numbers congruent to 1 mod 4 is also congruent to 1 mod 4, we have $N \equiv 1 \bmod 4$, a contradiction.

Section 3.5

1. The Euclidean algorithm gives $18(1) + 12(-1) = 6$, so $18(5) + 12(-5) = 30$. Thus $x = 5$ is a solution of $18x \equiv 30 \bmod 12$, or equivalently $3x \equiv 5 \bmod 2$. Thus the general solution is $x = 5 + 2u$, $y = -5 - 3u$. There are 6 distinct solutions $\bmod\ 12$, corresponding to $u = 0, 1, 2, 3, 4, 5$.

2. The Euclidean algorithm gives $11(-1) + 6(2) = 1$, so $x = -1$ is a solution of $11x \equiv 1 \bmod 6$, and we may replace -1 by 5 since $-1 \equiv 5 \bmod 6$. The general solution is $x = 5 + 6u$, $y = -9 - 11u$, which is unique $\bmod\ 6$.

3. The given equation is equivalent to $6x + 9y = 3$, and from the Euclidean algorithm we have $6(-1) + 9(1) = 3$, with $\gcd(6, 9) = 3$. Thus $6x \equiv 3 \bmod 9$ is equivalent to $2x \equiv 1 \bmod 3$, and $x = -1$, which can be replaced by $x = 2$, is a solution. The general solution is $x = 2 + 3u$, $y = -1 - 2u$. There are 3 distinct solutions of $6x \equiv 3 \bmod 9$, namely, $x = 2$, $x = 5$, and $x = 8$.

4. We have $m = 4(5)(9) = 180$, $y_1 = 180/4 = 45$, $y_2 = 180/5 = 36$, $y_3 = 180/9 = 20$. Since $45 \equiv 1 \bmod 4$, $36 \equiv 1 \bmod 5$, and $20 \equiv 2 \bmod 9$, we may take $z_1 = 1$, $z_2 = 1$, and $z_3 = 5$. Thus one solution is given by $x_0 = 2(45)(1) + 1(36)(1) + 6(20)(5) = 726 \equiv 6 \bmod 180$. The general solution is $x = 6 + 180u$, $u \in \mathbb{Z}$; the solution is unique $\bmod\ 180$.

5. If

$$\sum_{i=1}^{k} b_i y_i z_i \equiv 0 \bmod m$$

(hence $\bmod\ m_j$ for all j), then by (12) we have $b_j \equiv 0 \bmod m_j$ for all $j = 1, \ldots, k$. Now suppose that $(b_1, \ldots, b_k)$ and $(c_1, \ldots, c_k)$ both map to x_0. Since in (13), x_0 is a linear combination of the b_i, it follows that $(b_1 - c_1, \ldots, b_k - c_k)$ will map to $x_0 - x_0 = 0$. But then $b_i - c_i \equiv 0 \bmod m_i$, proving that the mapping is injective. Since $\mathbb{Z}_{m_1} \times \cdots \times \mathbb{Z}_{m_k}$ and $\mathbb{Z}_m$ each have m elements, the mapping is surjective by (1.5.2).

Section 3.6

1. (a) $600 = 2^3(3)(5^2)$, $\varphi(600) = 600(1 - \frac{1}{2})(1 - \frac{1}{3})(1 - \frac{1}{5}) = 160$
(b) $841 = 29^2$, $\varphi(841) = 29^2 - 29 = 812$
(c) $6174 = 2(3^2)(7^3)$, $\varphi(6174) = 6174(1 - \frac{1}{2})(1 - \frac{1}{3})(1 - \frac{1}{7}) = 1764$

2. The residues are 5, 1, 2, 7, 8, 4, a permutation of 1, 2, 4, 5, 7, 8.

3. Let $p_1, \ldots, p_r$ be the primes occurring in the factorization of m, and let $q_1, \ldots, q_s$ be the primes occurring in the factorization of n. Since m and n are relatively prime,

$p_i \neq q_j$ for all i, j. Thus

$$\varphi(mn) = mn \left(1 - \frac{1}{p_1}\right) \cdots \left(1 - \frac{1}{p_r}\right) \left(1 - \frac{1}{q_1}\right) \cdots \left(1 - \frac{1}{q_s}\right) = \varphi(m)\varphi(n).$$

4. Let $n = 4$, $r = 2$. Then $\binom{n}{r} = 6$, which is not divisible by 4.

5. If p does not divide a, then by Fermat's Theorem, the inverse of $a \bmod p$ is a^{p-2}. But for large p, the computation becomes very laborious.

6. Let $N = 2^{(p_1-1)\cdots(p_k-1)} = n + 1$. Since $p_1 > 2$, p_1 cannot divide 2 and therefore Fermat's Theorem implies that $2^{p_1-1} \equiv 1 \bmod p_1$. Successively raising both sides of this congruence to the powers $p_2 - 1, \ldots, p_k - 1$, we find that $N \equiv 1 \bmod p_1$. Since $p_2 > 2$, p_2 cannot divide 2^{p_1-1}, and Fermat's Theorem gives $2^{(p_1-1)(p_2-1)} \equiv 1 \bmod p_2$. As above, we conclude that $N \equiv 1 \bmod p_2$. Continuing in this fashion, we have $N \equiv 1 \bmod p_i$, $i = 1, 2, \ldots, k$. In other words, $n = N - 1$ is divisible by each p_i, and since the p_i are distinct primes, the product $p_1 \cdots p_k$ divides n (see (3.4.5 (f)).

Section 3.7

1. This follows because 17305893 is divisible by $9 = 3^2$.

2. If m is the product of r distinct primes and n is the product of s distinct primes, then since m and n are relatively prime, mn is the product of $r + s$ distinct primes. Thus $\mu(mn) = (-1)^{r+s} = (-1)^r(-1)^s = \mu(m)\mu(n)$. If m or n has a repeated prime factor, so does mn, and $\mu(mn) = \mu(m)\mu(n) = 0$.

3. A divisor d of n is of the form $d = p_1^{r_1} \ldots p_k^{r_k}$, $0 \leq r_i \leq e_i$. Since f is multiplicative, $f(d) = f(p_1^{r_1}) \ldots f(p_k^{r_k})$. Thus the terms in the expansion of

$$\left(1 + f(p_1) + \cdots + f(p_1^{e_1})\right) \ldots \left(1 + f(p_k) + \cdots + f(p_k^{e_k})\right)$$

correspond to the $f(d)$, $d|n$, and the result follows.

4. This follows from Problem 3, since $f(n) = n^r$ is multiplicative.

5. By definition, n is perfect if and only if the sum of all its positive divisors is $n + n = 2n$. Since $\sum_{d|n} d = S_1(n)$, the result follows.

6. $S_1(2^{n-1}) = 1 + 2 + 2^2 + \cdots + 2^{n-1} = 2^n - 1$, and since $2^n - 1$ is prime,

$$S_1(2^n - 1) = 1 + (2^n - 1) = 2^n.$$

Thus $S_1(x) = (2^n - 1)(2^n) = 2x$.

7. (a) By Problem 5, $S_1(x) = 2x$, and by Problem 4, $S_1(x) = S_1(2^h)S_1(q)$. But as in Problem 6, $S_1(2^h) = 2^{h+1} - 1$, and the result follows.

(b) $\dfrac{S_1(q)}{q} = \dfrac{2^{h+1}}{2^{h+1} - 1} > 1.$

(c) By (b), $2^{h+1}q = (2^{h+1}-1)S_1(q) = (2^{h+1}-1)(q+r)$, so $0 = -q+(2^{h+1}-1)r$, as asserted.

(d) If $r > 1$, then r is a divisor of q and $1 < r < q$. Thus $S_1(q) \geq q+r+1 > q+r$, contradicting (c).

(e) By (c) and (d), $S_1(q) = q + 1$, so the only positive divisors of q are q and 1. It follows that q must be prime.

Section 4.1

1. $\{1, 2, 4, 12\}$ and $\{1, 2, 6, 12\}$

2. If you visualize the ordered pair (a, b) as determined by a vertical line (column) at $x = a$ and a horizontal line (row) at $y = b$ in an x-y plane, then to compare two pairs, we first look at columns, and if the columns are equal, we then look at rows. It should be clear intuitively that we have a total ordering, and the formal details are straightforward. Let C be a nonempty subset of $A \times B$. Among all first coordinates a of ordered pairs $(a, b) \in C$, there is a smallest element a_0, and among all second coordinates b of ordered pairs $(a_0, b) \in C$, there is a smallest element b_0. If $(a, b) \in C$, then

Case 1: $a_0 < a$. Then $(a_0, b_0) < (a, b)$.
Case 2: $a_0 = a$. Then $(a_0, b) = (a, b) \in C$, so $(a_0, b_0) \leq (a_0, b) = (a, b)$.

Thus (a_0, b_0) is the smallest element of C.

3. No, the ordering is not even total, assuming that A and B each have at least two elements. For if $a_1 < a_2$ and $b_1 < b_2$, then (a_1, b_2) and (a_2, b_1) cannot be compared.

Section 4.2

1. Assuming Zorn's Lemma, let B be a chain of the partially ordered set A. The collection C of all chains containing B is nonempty (since B is a chain containing B) and is partially ordered by inclusion. (See Section 1.6, Problem 7.) Every chain of C has an upper bound in C, namely the union of all the chains of A that comprise the chain of C. By Zorn's Lemma, there is a maximal element, in other words, a maximal chain containing B.

2. If $r_1 v_1 + \cdots + r_n v_n = 0$ but not all $r_i = 0$, say $r_1 \neq 0$. Then

$$v_1 = -r_1^{-1} r_2 v_2 - r_1^{-1} r_3 v_3 - \cdots - r_1^{-1} r_n v_n,$$

so that v_1 can be expressed as a linear combination of $v_2, \ldots, v_n$. Conversely, if one of the v_i can be expressed as a linear combination of the others, move all v_i to the same side of the equation to conclude that a nontrivial linear combination of the v_i is 0.

3. The argument is virtually identical to that of Problem 2.

4. Suppose that the chain consists of the linearly independent sets L_i, $i \in I$. Then each L_i is contained in the union of all the L_i, so $\bigcup_{i \in I} L_i$ is an upper bound of the chain in $\mathcal{C}$, provided we can show that it is a linearly independent set. But if $r_1 v_1 + \cdots + r_n v_n = 0$ with the $v_j \in \bigcup_{i \in I} L_i$, then for some index k we have all $v_j \in L_k$, because the L_i form a chain. (For example, if $v_1 \in L_1$, $v_2 \in L_7$, and $L_1 \subseteq L_7$, then both v_1 and v_2 belong to L_7.) Since L_k is linearly independent, all r_i must be 0.

5. By Zorn's Lemma, $\mathcal{C}$ has a maximal element, that is, V has a maximal linearly independent set.

Section 4.3

1. If $c \in C$, there is an element $b \in B$ such that $g(b) = c$, and an element $a \in A$ such that $f(a) = b$. But then $g(f(a)) = c$, proving $g \in f$ surjective.

2. There are many possibilities. For example, let $f(x) = x^2$ on the reals. Then f is not injective, but if we restrict f to the nonnegative reals, the resulting function is injective.

3. If $B \leq_s A$, then there is an injective map g from B into A. The inverse of this function maps $g(B)$ onto B. If we define $f(x) = g(x)$ for $x \in g(B)$, and define $f(x)$ to be an arbitrary element of B for $x \in A \setminus g(B)$, then $f : A \to B$ is surjective. Conversely, if f maps A onto B, then for each $y \in B$ there is an element $x \in A$ such that $f(x) = y$. Choose one such x (Axiom of Choice!) and call it $g(y)$. If $x = g(y_1) = g(y_2)$, then by definition of g, x is mapped by f to both y_1 and y_2, and since f is a function, we must conclude that $y_1 = y_2$. Thus g is an injective map of B into A, so $B \leq_s A$.

4. If B is countably infinite, then there is a bijection between B and $\mathbb{N}$, and if B is finite, there is an injective map from B to $\mathbb{N}$. Thus B is countable if and only if $B \leq_s \mathbb{N}$, and the result follows from Problem 3.

5. Let $A = \{a_1, \ldots, a_m\}$ and $B = \{b_1, \ldots, b_n\}$. If $m = n$, then $a_i \to b_i$, $1 \leq i \leq n$, defines a bijection between A and B, so $|A| = |B|$. If $m < n$, then $a_i \to b_i$, $1 \leq i \leq m$, defines an injective map from A to B, so $|A| \leq |B|$. Since $m < n$, any function from B to A must map at least two b_i's to the same element of A, so there can be no bijection between A and B. Therefore $|A| < |B|$, and the result follows.

Section 4.4

In Problems 1, 2, and 3, $|A| = \alpha$, $|B| = \beta$, $|C| = \gamma$.

1. A^{B+C} is the set of functions from the disjoint union of B and C to A, and this set of functions is in one-to-one correspondence with the set of pairs of functions (f, g) where $f : B \to A$ and $g : C \to A$. (If $h : B + C \to A$, take f and g to be the restrictions of h to B and C, respectively.) Thus $|A^{B+C}| = |A^B||A^C|$.

2. $(A \times B)^C$ is the set of functions $f : C \to A \times B$, and f corresponds to a pair (g, h) with $g : C \to A$ and $h : C \to B$. Explicitly, if $f(c) = (a, b)$ then $g(c) = a$ and $h(c) = b$. Thus $(A \times B)^C$ has the same cardinality as $A^C B^C$.

3. If $f : B \times C \to A$, define $f_c : B \to A$ as $f_c(b) = f(b, c)$. Then f determines a mapping $c \to f_c$ from C to A^B. Conversely, given the mapping $c \to f_c$, we can recapture f by $f(b, c) = f_c(b)$. This establishes a one-to-one correspondence between $A^{B \times C}$ and $(A^B)^C$.

4. If B is any infinite set, then B has a countably infinite subset C, as we found in the proof of (4.4.3(b)). Thus $\aleph_0 = |C| \leq |B|$.

5. A real number may be specified by selecting an interval $[n, n+1)$ and then choosing a point in that interval. If α is the cardinality of the set of reals between 0 and 1, then each interval $[n, n+1)$ has cardinality α, so $c = \aleph_0 \alpha$. But $\aleph_0 < \alpha$ (see Section 2.4), and consequently $c = \alpha$ by (4.4.3(b)).

6. An element of A can be identified with a finite subset of the positive integers. For example, 01001 has 1's in positions 2 and 5, and therefore corresponds to $\{2, 5\}$. But we know that there are only countably many finite subsets of the positive integers (see (2.4.3) and the discussion preceding it, or (4.4.4)).

7. By Problems 5 and 6, $2^{\aleph_0} = c + \aleph_0$, and since $\aleph_0 < c$, we have $c + \aleph_0 = c$ by (4.4.3(a)).

8. By (4.3.7), $2^{\aleph_0} > \aleph_0$, and since $\aleph_1$ is the smallest cardinal greater than $\aleph_0$, we must have $\aleph_1 \leq 2^{\aleph_0}$.

Section 5.1

1. The ij element of $A(B+C)$ is $\sum_k a_{ik}(b_{kj} + c_{kj}) = \sum_k a_{ik} b_{kj} + \sum_k a_{ik} c_{kj}$, which is the ij element of AB plus the ij element of AC. The second distributive law is proved similarly. The key point is that the distributive laws hold for real numbers, in fact for any field.

2. The ij element of $A(BC)$ is

$$\sum_k a_{ik} \sum_r b_{kr} c_{rj} = \sum_r \left(\sum_k a_{ik} b_{kr} \right) c_{rj} = \sum_r (AB)_{ir} C_{rj},$$

which is the ij element of $(AB)C$. The key points are that multiplication is associative in any field, and the order of summation of a finite double series can always be reversed.

3. No. For example, let

$$A = \begin{bmatrix} 0 & 1 \\ 0 & 0 \end{bmatrix}, \quad B = \begin{bmatrix} 1 & 0 \\ 0 & 0 \end{bmatrix}, \quad C = \begin{bmatrix} 0 & 0 \\ 0 & 0 \end{bmatrix}.$$

4. If we apply the row operations represented by $E_1, \ldots, E_k$ to A in that order, the result is the product $E_k E_{k-1} \cdots E_1 A$, which is I_n by hypothesis. But if the row operations are applied to I_n, we get $E_k E_{k-1} \cdots E_1 I_n = E_k E_{k-1} \cdots E_1 = B$. Therefore $BA = I_n$.

5. Multiply $BA = I_n$ on the right by A^{-1} to obtain $B = A^{-1}$.

6. A is an elementary row matrix obtained from I_2 by adding 3 times row 2 to row 1. Thus A^2 is obtained from A by adding 3 times row 2 to row 1; the result is

$$\begin{bmatrix} 1 & 6 \\ 0 & 1 \end{bmatrix}.$$

Continuing in this fashion, we have

$$A^k = \begin{bmatrix} 1 & 3k \\ 0 & 1 \end{bmatrix};$$

in particular,

$$A^{74} = \begin{bmatrix} 1 & 222 \\ 0 & 1 \end{bmatrix}.$$

7. If A is $m \times n$, then A^t is $n \times m$, so that AA^t exists and is $m \times m$. Since $(AA^t)^t = (A^t)^t A^t = AA^t$, it follows that AA^t is symmetric.

8. No. As in the text we have $a_{ii} = -a_{ii}$ so $a_{ii} + a_{ii} = 0$. In a field of "characteristic 2", in other words a field in which $1 + 1 = 0$, it does not follow that $a_{ii} = 0$. We have already met one such field, namely $\mathbb{Z}_2$, the field of integers modulo 2.

9. We have $A = \frac{1}{2}(A + A^t) + \frac{1}{2}(A - A^t) =$ symmetric + skew-symmetric.

10. Direct computation shows that A^2 has 1's in the 1-3 and 2-4 positions, and 0's elsewhere; A^3 has a 1 in the 1-4 position, and 0's elsewhere; A^4 has all zero entries.

Section 5.2

1. Apply the elementary row operations $R_2 \leftarrow R_2 - 2R_1$, $R_2 \leftarrow -\frac{1}{5}R_2$, $R_1 \leftarrow R_1 - 3R_2$, $R_3 \leftarrow R_3 - R_2$, $R_3 \leftarrow -R_3$ to I_3 to get

$$A^{-1} = \begin{bmatrix} -1/5 & 3/5 & 0 \\ 2/5 & -1/5 & 0 \\ 2/5 & -1/5 & -1 \end{bmatrix}.$$

2. In Problem 1, the second operation multiplies the determinant by $-\frac{1}{5}$, and the fifth operation multiplies the determinant by -1; the other operations leave the determinant unchanged. Thus $\det A = \frac{1}{1/5} = 5$. Checking by Laplace Expansion down column 3, we have $(-1)(1 - 6) = 5$.

3. If rows i and j are identical, add -1 times row i to row j to produce a row of zeros, and therefore a zero determinant.

4. If $R_i = a_1 R_{i_1} + \cdots + a_k R_{i_k}$, successively add $-a_1$ times row $i_1, \ldots, -a_k$ times row i_k to row i to produce a row of zeros, and therefore a zero determinant.

5. If a sequence of elementary row operations reduces A to echelon form Q, then the analogous sequence of elementary column operations will reduce A^t to Q^t. (If $BA = Q$, then $A^t B^t = Q^t$.) If $Q = I$, then $Q^t = I$, and if Q has a row of zeros, then Q^t has a column of zeros. Thus the computational procedure for finding the determinant of A^t produces exactly the same set of numbers as the procedure for finding the determinant of A. Therefore $\det A^t = \det A$.

Section 5.3

1. If S is a basis then S spans, so each $x \in V$ has an expression of the desired form. If x has two distinct representations then $u_1, \ldots, u_n$ are linearly dependent, a contradiction. Conversely, if each x is a linear combination of the u_i, then S spans V. If $a_1 u_1 + \cdots + a_n u_n = 0$, then since 0 has the unique representation $0u_1 + \cdots + 0u_n$, we have $a_1 = \cdots = a_n = 0$.

2. Lining up u, v, and w as columns, we have

$$A = \begin{bmatrix} 1 & 0 & 0 \\ 0 & 1 & 0 \\ 0 & 1 & 1 \end{bmatrix}.$$

Since the echelon form of A is I_3 (equivalently, A is invertible; equivalently, $\det A \neq 0$), the equations $au + bv + cw = 0$ have the unique solution $a = b = c = 0$. Therefore u, v, and w are three linearly independent vectors in $\mathbb{R}^3$, hence a basis.

3. With A as in Problem 2, we must solve the equations

$$A \begin{bmatrix} a \\ b \\ c \end{bmatrix} = \begin{bmatrix} 2 \\ 3 \\ 4 \end{bmatrix}$$

for a, b, and c. The result is $a = 2$, $b = 3$, $c = 1$.

4. Assume that one of the bases, say T, is finite. The proof of (5.3.2) applies verbatim, and shows that $|S| \leq |T|$. But then S is also finite.

5. If $j \in I$ but j does not belong to the union of the $I(x)$, then for any $x \in S$, x depends on the y_i, $i \in I(x)$, but i is never equal to j. Thus the vectors in S can be expressed in terms of $T \setminus \{y_j\}$, a contradiction since T is a basis, hence a minimal spanning set.

6. An element of $\cup\{I(x) : x \in S\}$ is determined by selecting a vector $x \in S$ and then choosing an index in $I(x)$. Since $I(x)$ is finite, we have $|I(x)| \leq \aleph_0$. By Problem 5, we have $|I| = |\cup\{I(x) : x \in S\}|$, so $|I| \leq |S|\aleph_0$, and the result follows.

Section 5.4

1.   (a) The first quadrant $\{(x, y) : x \geq 0 \text{ and } y \geq 0\}$.

    (b) The union of the first quadrant and the third quadrant $\{(x, y) : x \leq 0 \text{ and } y \leq 0\}$.

2. The fourth component of a vector in S is twice the first component minus the second component. This property is maintained under addition and scalar multiplication, so S is a subspace.

3. Let $a = 1$, $b = c = 0$ to get $u = (1, 0, 0, 2)$; let $a = 0$, $b = 1$, $c = 0$ to get $v = (0, 1, 0, -1)$; let $a = b = 0$, $c = 1$ to get $w = (0, 0, 1, 0)$. Our choices of a, b, and c guarantee that u, v, w are linearly independent. If $p = (a, b, c, 2a - b)$ is any vector in S, then $p = au + bv + cw$. Thus u, v, and w span and therefore form a basis.

4. If $a(u + v) + b(v + w) + c(w + u) = 0$, then by linear independence of u, v, w we have $a + c = 0$, $a + b = 0$, and $b + c = 0$. These equations have a unique solution $a = b = c = 0$, so $u + v$, $v + w$, and $w + u$ are three linearly independent vectors in a three-dimensional subspace. Thus $u + v$, $v + w$, and $w + u$ are a basis.

5. Line up the vectors as columns to obtain

$$\begin{bmatrix} 1 & 2 & 0 & 1 \\ 0 & 1 & 1 & 4 \\ 2 & 3 & 1 & 8 \end{bmatrix}.$$

Elementary row operations yield the echelon form

$$\begin{bmatrix} 1 & 0 & 0 & 3 \\ 0 & 1 & 0 & -1 \\ 0 & 0 & 1 & 5 \end{bmatrix}.$$

If C_i is column i, then C_1, C_2, and C_3 are linearly independent, and it follows that u, v, and w are a basis. Since $C_4 = 3C_1 - C_2 + 5C_3$, we have $(1, 4, 8) = 3u - v + 5w$.

6. (a) K will be a subspace, typically a line or a plane through the origin. Then C will be a translated subspace, in other words, a line or a plane not necessarily through the origin.

(b) Suppose $u+K = v+K$. Then $u = u+0 \in u+K = v+K$, so $u-v \in K = N(A)$. But then $A(u-v) = 0$, hence $Au = Av$. Note also that if $u-v \in K$, then $u+K = v+K$, for if $w \in u + K$, then $w = u + p$, $p \in K$, and also $u = v + q$, $q \in K$. Thus $w = u + p = v + (p + q) \in v + K$, so $u + K \subseteq v + K$; the reverse inclusion is proved by a symmetrical argument. This observation will be useful in Problem 7.

7. (a) If $u_1+K = u_2+K$ and $v_1+K = v_2+K$, then u_1-u_2 and v_1-v_2 belong to K, so $(u_1-u_2)+(v_1-v_2) = (u_1+v_1)-(u_2+v_2) \in K$. Therefore $(u_1+v_1)+K = (u_2+v_2)+K$. Similarly $au_1 - au_2 = a(u_1 - u_2) \in K$, so $au_1 + K = au_2 + K$.

(b) $\pi\big(a(u + K) + b(v + K)\big) = \pi(au + bv + K) = A(au + bv) = aAu + bAv$
$$= a\pi(u + K) + b\pi(v + K).$$

(c) If $\pi(u + K) = \pi(v + K)$, then $Au = Av$, so $A(u - v) = 0$, and therefore $u - v \in K$. But then $u + K = v + K$, proving that π is injective. Since $\pi(u+K) = Au$, which ranges over all of $R(A)$ as u ranges over F^n, π is surjective.

Section 5.5

1. If $u = Tx$ and $v = Ty$, then $u + v = Tx + Ty = T(x + y)$, so $T^{-1}(u + v) = x + y = T^{-1}u + T^{-1}v$. Also, $cu = cTx = T(cx)$, so $T^{-1}(cu) = cx = cT^{-1}u$, proving that T^{-1} is linear. If the matrix B represents T^{-1} then since $T^{-1} \circ T$ is the identity transformation, represented by the identity matrix I, we have $BA = I$, so $B = A^{-1}$.

2. New coordinates $= P^{-1}$(old coordinates), so

$$P^{-1} = \begin{bmatrix} 4 & -6 \\ 0 & 1 \end{bmatrix} \quad \text{and} \quad P = \begin{bmatrix} 1/4 & 3/2 \\ 0 & 1 \end{bmatrix}.$$

Therefore

$$u = \begin{bmatrix} 1/4 \\ 0 \end{bmatrix} = \frac{1}{4}e_1 \quad \text{and} \quad v = \begin{bmatrix} 3/2 \\ 1 \end{bmatrix} = \frac{3}{2}e_1 + e_2.$$

3. $T(1, 0)$ has length 1 and angle θ, so $T(1, 0) = (\cos\theta, \sin\theta)$. $T(0, 1)$ has length 1 and angle $\frac{\pi}{2} + \theta$, so $T(0, 1) = (\cos(\frac{\pi}{2} + \theta), \sin(\frac{\pi}{2} + \theta)) = (-\sin\theta, \cos\theta)$. The matrix of T with respect to the standard basis is

$$A = \begin{bmatrix} \cos\theta & -\sin\theta \\ \sin\theta & \cos\theta \end{bmatrix}.$$

4. $T(u) = u$ and $T(v) = -v$, so $B = \begin{bmatrix} 1 & 0 \\ 0 & -1 \end{bmatrix}$. The basis-changing matrix is $P = \begin{bmatrix} 1 & -a \\ a & 1 \end{bmatrix}$. Then

$$P^{-1} = \frac{1}{1+a^2}\begin{bmatrix} 1 & a \\ -a & 1 \end{bmatrix} \quad \text{and} \quad B = P^{-1}AP.$$

Thus

$$A = PBP^{-1} = \frac{1}{1+a^2}\begin{bmatrix} 1-a^2 & 2a \\ 2a & a^2-1 \end{bmatrix}.$$

5. If T is a linear transformation represented by the matrix A with respect to a given basis, the mapping $x \to Tx$ corresponds to the matrix calculation $c \to Ac$. The image of T corresponds to $R(A)$, the range of A. If the basis is changed, the same linear transformation T is represented by a matrix B similar to A, and now the image of T corresponds to $R(B)$. Therefore $R(A)$ and $R(B)$ have the same dimension, that is, rank A = rank B.

6. If $B = P^{-1}AP$, then $B^t = P^t A^t (P^{-1})^t = P^t A^t (P^t)^{-1} = Q^{-1} A^t Q$, where $Q = (P^t)^{-1}$.

7. Both results follow from (5.5.5): $\dim(\ker T) + \dim(\operatorname{im} T) = \dim V$.

(a) If $\ker T = \{0\}$, then $\dim(\operatorname{im} T) = \dim V > \dim W$, which is impossible since $\operatorname{im} T \subseteq W$. Thus $\ker T$ contains a nonzero vector, so T is not injective.

(b) If $\operatorname{im} T = W$, then $\dim(\ker T) = \dim V - \dim W < 0$, a contradiction. Thus the image of T must be a proper subset of W, so that T is not surjective.

Section 5.6

1. $\|x+y\|^2 = \|x\|^2 + \|y\|^2 + 2\operatorname{Re}\langle x, y\rangle;$
$\|x-y\|^2 = \|x\|^2 + \|y\|^2 - 2\operatorname{Re}\langle x, y\rangle;$
$\|x+iy\|^2 = \|x\|^2 + \|y\|^2 + 2\operatorname{Re}\langle x, iy\rangle;$
$\|x-iy\|^2 = \|x\|^2 + \|y\|^2 - 2\operatorname{Re}\langle x, iy\rangle.$

But $\operatorname{Re}\langle x, iy\rangle = \operatorname{Re}[-i\langle x, y\rangle] = \operatorname{Im}\langle x, y\rangle$, and the result follows.

2. This follows from the last equation in the proof of (5.6.7), with $a_i = \langle x, x_i\rangle$.

3. If $z \in S$ and $x, y \in S^\perp$, $a, b \in C$, then $\langle ax + by, z\rangle = a\langle x, z\rangle + b\langle y, z\rangle = 0$. Thus $S^\perp$ is closed under linear combination and is therefore a subspace.

4. By the Projection Theorem (5.6.9), p is the unique vector in S such that $x - p$ is orthogonal to each x_i. Since the components of x_i will appear in row i of A^t, we have $A^t(x - p) = 0$, or $A^t x = A^t p$. But $p = a_1 x_1 + \cdots + a_k x_k = a_1$ (column 1 of A) $+ \cdots + a_k$ (column k of A) $= Aq$, as can be visualized by walking across a row of A and down the column vector q. Thus $A^t x = A^t Aq$. If the scalars are allowed to be complex, the normal equations become $A^* x = A^* Aq$, where A^* is the conjugate transpose of A; that is, A^* is formed by taking the complex conjugate of each element of A, and then transposing. (The condition that $x - p$ is orthogonal to each x_i can be expressed as $A^*(x - p) = 0$; the remainder of the analysis is the same.)

5. We have $E = \|Y - AX\|^2$, and as the components of X range over all real numbers, the vectors AX range over the space spanned by the columns of A. Thus we are projecting Y on the space spanned by the columns of A. The result follows from Problem 4.

6. The vector Y is the same as in Problem 5, but now we have

$$E = \sum_{i=1}^{m} |y_i - ax_i^2 - bx_i - c|^2 \quad \text{and} \quad X = \begin{bmatrix} a \\ b \\ c \end{bmatrix}.$$

The matrix A now has three columns. The components of the first column are $x_1^2, \ldots, x_m^2$, the components of the second column are $x_1, \ldots, x_m$, and the components of the third column are $1, \ldots, 1$.

7. Equality holds if and only if x and y are linearly dependent. For if there is equality, then by the proof of (5.6.2), $x + ay = 0$ for some a. (If $y = 0$, then equality holds, and x and y are linearly dependent as well, so this case causes no difficulty.) Conversely, if x and y are linearly dependent, then one is a multiple of the other, say $x = cy$. Then

$$|\langle x, y \rangle| = |\langle cy, y \rangle| = |c| \, \|y\|^2 = (|c| \|y\|) \|y\| = \|x\| \, \|y\|.$$

Section 5.7

1. If A and B are unitary, then $(AB)(AB)^* = ABB^* A^* = AIA^* = AA^* = I$, proving that AB is unitary. The sum need not be unitary; for example, take $B = -A$.

2. If $Tx = \lambda x$, then $T^2 x = T(Tx) = T(\lambda x) = \lambda(Tx) = \lambda(\lambda x) = \lambda^2 x$. Apply T successively to get the result.

3. $\det(A - \lambda I) = (2 - \lambda)^2(1 - \lambda)$, so the eigenvalues are $\lambda = 2$ (2-fold) and $\lambda = 1$. When $\lambda = 2$, the equations

$$(A - \lambda I) \begin{bmatrix} x \\ y \\ z \end{bmatrix} = 0$$

become $y = 0$, $z = 0$, x arbitrary. The eigenspace is only one-dimensional, spanned by $(1, 0, 0)$. When $\lambda = 1$, the equations are $x + y = 0$, $y = 0$, z arbitrary, so the eigenspace is spanned by $(0, 0, 1)$. There are only two linearly independent eigenvectors in a three-dimensional space, so A cannot be diagonalized.

4. A is invertible if and only if $\det A = \det(A - 0I) \neq 0$, in other words, 0 is not an eigenvalue of A.

5. $(2, 4)$ and $(-7, y)$ are orthogonal by (5.7.7), so $-14 + 4y = 0$, $y = 7/2$.

6. We have $A = UDU^*$, so $A^* = U^{**}D^*U^* = UDU^* = A$.

7. If A is similar to the matrix $D = \operatorname{diag}(\lambda_1, \ldots, \lambda_n)$, then by (5.7.2), $\det A = \det D = \lambda_1 \ldots \lambda_n$.

8. $A^2 = PDP^{-1}PDP^{-1} = PD^2P^{-1}$, and by iteration, $A^k = PD^kP^{-1}$. But D^k is a diagonal matrix with entries $\lambda_1^k, \ldots, \lambda_n^k$, so A^k is relatively easy to compute.

9. $q = 3(x^2 + \frac{2}{3}xy + \frac{1}{9}y^2) - y^2 - \frac{1}{3}y^2 = 3(x + \frac{1}{3}y)^2 - \frac{4}{3}y^2 = 3X^2 - \frac{4}{3}Y^2$ where $X = x + \frac{1}{3}y$, $Y = y$. Thus

$$\begin{bmatrix} X \\ Y \end{bmatrix} = \begin{bmatrix} 1 & 1/3 \\ 0 & 1 \end{bmatrix} \begin{bmatrix} x \\ y \end{bmatrix},$$

and by (5.5.6),

$$P^{-1} = \begin{bmatrix} 1 & 1/3 \\ 0 & 1 \end{bmatrix}.$$

Invert P^{-1} to get

$$P = \begin{bmatrix} 1 & -1/3 \\ 0 & 1 \end{bmatrix}.$$

The new basis vectors are $(1, 0)$ and $(-1/3, 1)$.

10. $q = 3(x^2 + (2y + 6z)x + (y + 3z)^2) - 6y^2 + z^2 - 3(y + 3z)^2$
        $= 3(x + y + 3z)^2 - 9y^2 - 18yz - 26z^2$;

then proceed to reduce $-9y^2 - 18yz - 26z^2$ as in Problem 9.

11. $\|Ux\|^2 = \langle Ux, Ux \rangle = (Ux)^*Ux = x^*U^*Ux = x^*x = \langle x, x \rangle = \|x\|^2$.

12. Let x be an eigenvector for λ. Then $Ux = \lambda x$, and by Problem 11, $\|Ux\| = \|x\|$, so $\|x\| = \|\lambda x\| = |\lambda| \, \|x\|$. Therefore $|\lambda| = 1$.

Section 6.1

1. The largest Jordan block has order 3, and in fact there are 2 blocks of order 3. Since $\text{rank}(J - \lambda I) = 2$ (# of blocks of order 3) $+ 1$ (# of blocks of order 2), there are $7 - 4 = 3$ blocks of order 2. The # of blocks of order 1 is

$$14 - 3 \text{ (\# of blocks of order 3)} - 2 \text{ (\# of blocks of order 2)} = 14 - 6 - 6 = 2.$$

2. The largest Jordan block must have order 3, and there must be only 1 block of this order. Therefore the conditions are

$$\text{rank}(J - \lambda I)^3 = 0, \qquad \text{rank}(J - \lambda I)^2 = 1$$

3. In this case, the rank of $J - \lambda I$ must be 0, in other words, $J - \lambda I$ must be the zero matrix.

4. Look at the 18 by 18 matrix J at the beginning of the section. The determinant of J is 3^{18}, and since $\det(J - \lambda I) = (3 - \lambda)^{18}$, the multiplicity of the eigenvalue 3 is 18. This argument works in the general case, and the result now follows from the fact that the Jordan canonical form is a direct sum of matrices $J(\lambda)$, λ ranging over all eigenvalues of A.

Section 6.2

1. J is already in Jordan canonical form, and its characteristic polynomial is $c(x) = (x - \lambda)^r$. Thus J has only one eigenvalue λ, of multiplicity r. In this case, there is only one Jordan block, of order r. By (6.2.4), the minimal polynomial of J is $m(x) = (x - \lambda)^r$.

2. By (6.2.6), $c(x) = (x - \lambda_1) \cdots (x - \lambda_n)$. By (5.7.4), A is diagonalizable, so by (6.2.4) and (6.2.5), $m(x)$ coincides with $c(x)$. The Jordan canonical form is $\text{diag}(\lambda_1, \ldots, \lambda_n)$.

3. Case 1: $m(x) = c(x)$. Then corresponding to λ_1 there is one Jordan block of order 2, and corresponding to λ_2 there is one Jordan block of order 1. The Jordan canonical form is

$$\begin{bmatrix} \lambda_1 & 1 & 0 \\ 0 & \lambda_1 & 0 \\ 0 & 0 & \lambda_2 \end{bmatrix}.$$

Case 2: $m(x) = (x - \lambda_1)(x - \lambda_2)$. Then corresponding to λ_1 there are two blocks of order 1, and corresponding to λ_2 there is one block of order 1. The Jordan canonical form is

$$\begin{bmatrix} \lambda_1 & 0 & 0 \\ 0 & \lambda_1 & 0 \\ 0 & 0 & \lambda_2 \end{bmatrix}.$$

4. Case 1: $m(x) = c(x)$. Then there is only one Jordan block, of order 3, and the Jordan canonical form is

$$\begin{bmatrix} \lambda & 1 & 0 \\ 0 & \lambda & 1 \\ 0 & 0 & \lambda \end{bmatrix}.$$

Case 2: $m(x) = (x - \lambda)^2$. There is one block of order 2 and one block of order 1, and the Jordan canonical form is

$$\begin{bmatrix} \lambda & 1 & 0 \\ 0 & \lambda & 0 \\ 0 & 0 & \lambda \end{bmatrix}.$$

Case 3: $m(x) = x - \lambda$. There are three blocks of order 1, and the Jordan canonical form is

$$\begin{bmatrix} \lambda & 0 & 0 \\ 0 & \lambda & 0 \\ 0 & 0 & \lambda \end{bmatrix}.$$

5. Let A be a 4 by 4 matrix with characteristic polynomial $c(x) = (x - \lambda)^4$ and minimal polynomial $m(x) = (x - \lambda)^2$. Then the largest block is of order 2, giving rise to a submatrix

$$\begin{bmatrix} \lambda & 1 \\ 0 & \lambda \end{bmatrix}.$$

There can be another Jordan block of order 2, or two blocks of order 1, so the Jordan form is not determined by simply giving $c(x)$ and $m(x)$.

6. Suppose that $c(x) = \sum_{i=0}^{n} a_i x^i$; then $\sum_{i=0}^{n} a_i A^i = 0$ by Cayley-Hamilton (take A^0 to be I). If A is known to be invertible, we can multiply both sides of the equation by A^{-1} to get $a_0 A^{-1} + \sum_{i=1}^{n} a_i A^{i-1} = 0$, so that A^{-1} can be expressed in terms of powers of A. Notice that if $a_0 = 0$, then x is a factor of $c(x)$, so that 0 is an eigenvalue of A. But then A can't be invertible (see Section 5.7, Problem 4).

Section 6.3

1. $\langle x, (T + S)^* y \rangle = \langle (T + S)x, y \rangle = \langle Tx + Sx, y \rangle = \langle Tx, y \rangle + \langle Sx, y \rangle$
$\qquad\qquad = \langle x, T^* y \rangle + \langle x, S^* y \rangle = \langle x, T^* y + S^* y \rangle,$
so $(T + S)^* y = T^* y + S^* y$, that is, $(T + S)^* = T^* + S^*$.

2. $\langle x, (cT)^* y \rangle = \langle (cT)x, y \rangle = \langle cTx, y \rangle = c \langle Tx, y \rangle = c \langle x, T^* y \rangle$
$\qquad\qquad = \langle x, \overline{c} T^* y \rangle$, so $(cT)^* = \overline{c} T^*$.

3. $\langle x, (TS)^* y \rangle = \langle TSx, y \rangle = \langle Sx, T^* y \rangle = \langle x, S^* T^* y \rangle$, so $(TS)^* = S^* T^*$.

4. $\langle Tx, y \rangle = \langle x, T^* y \rangle = \overline{\langle T^* y, x \rangle} = \overline{\langle y, T^{**} x \rangle} = \langle T^{**} x, y \rangle$, so $T^{**} = T$.

5. $\langle x, I^*y \rangle = \langle Ix, y \rangle = \langle x, y \rangle$, so $I^* = I$.

6. $Tx = 0$ iff $\langle Tx, y \rangle = 0$ for all y iff $\langle x, T^*y \rangle = 0$ for all y.

7. By Problem 6, the kernel of T^* and the image of T^{**} are orthogonal complements. But by Problem 4, $T^{**} = T$ and the result follows.

Section 6.4

1. A has distinct eigenvalues $\lambda = 1$ and $\lambda = 2$, so A is diagonalizable. But $AA^* \neq A^*A$, so A is not unitarily diagonalizable.

2. Take $g = \sum_{i=0}^{n} b_i f_i$.

3. Since the T_i are projection operators, this is immediate from (6.3.7).

4. We have $T^2 = (\lambda_1 T_1 + \cdots + \lambda_k T_k)(\lambda_1 T_1 + \cdots + \lambda_k T_k) = \lambda_1^2 T_1 + \cdots + \lambda_k^2 T_k$, and similarly $T^m = \sum_{i=1}^{k} \lambda_i^m T_i$ for all m. Thus

$$a_0 I + a_1 T + \cdots + a_n T^n$$

$$= (a_0 + a_1\lambda_1 + \cdots + a_n\lambda_1^n)T_1 + \cdots + (a_0 + a_1\lambda_k + \cdots + a_n\lambda_k^n)T_k,$$

and the result follows.

5. If $T^* = g(T)$, then $TT^* = Tg(T) = g(T)T = T^*T$, so T is normal. If T is normal, write $T = \lambda_1 T_1 + \cdots + \lambda_k T_k$ as in (6.4.5). By (6.3.5), $T^* = \overline{\lambda_1}T_1^* + \cdots + \overline{\lambda_k}T_k^* = \overline{\lambda_1}T_1 + \cdots + \overline{\lambda_k}T_k$ by Problem 3. By Problem 2, there is a polynomial g such that $g(\lambda_i) = \overline{\lambda_i}$, $i = 1, \ldots, k$. Thus $T^* = g(\lambda_1)T_1 + \cdots + g(\lambda_k)T_k = g(T)$ by Problem 4.

6. If T is unitary, then T is normal by (6.4.1), and the eigenvalues of T have magnitude 1 by Section 5.7, Problem 12. Conversely, assume T normal with $|\lambda| = 1$ for all eigenvalues λ. Then by (6.4.5) and (6.3.5),

$$TT^* = (\lambda_1 T_1 + \cdots + \lambda_k T_k)(\overline{\lambda_1}T_1^* + \cdots + \overline{\lambda_k}T_k^*)$$

$$= (\lambda_1 T_1 + \cdots + \lambda_k T_k)(\overline{\lambda_1}T_1 + \cdots + \overline{\lambda_k}T_k) \quad \text{by Problem 3}$$

$$= |\lambda_1|^2 T_1 + \cdots + |\lambda_k|^2 T_k = T_1 + \cdots + T_k = I \quad \text{by (6.4.5)},$$

proving T unitary.

7. If T is self-adjoint then all eigenvalues of T are real by (5.7.7). Conversely, assume that all eigenvalues of T are real. Then $T = \lambda_1 T_1 + \cdots + \lambda_k T_k$ and

$$T^* = \overline{\lambda_1}T_1^* + \cdots + \overline{\lambda_k}T_k^* \quad \text{by (6.3.5)}$$

$$= \overline{\lambda_1}T_1 + \cdots + \overline{\lambda_k}T_k \quad \text{by Problem 3}$$

$$= \lambda_1 T_1 + \cdots + +\lambda_k T_k \quad \text{since the } \lambda_i \text{ are real.}$$

Thus $T^* = T$, so that T is self-adjoint.

8. For each $i = 1, \ldots, k$, let f_i be a polynomial such that

$$f_i(\lambda_j) = \delta_{ij} = \begin{cases} 0, & i \neq j \\ 1, & i = j \end{cases}$$

(see Problem 2). By Problem 4,

$$\begin{aligned} f_i(T) &= f_i(\lambda_1)T_1 + \cdots + f_i(\lambda_k)T_k \\ &= \delta_{i1}T_1 + \cdots + \delta_{ik}T_k = T_i. \end{aligned}$$

9. By (6.3.5) and Problems 3 and 4, $f(T)^*$ is a linear combination of the T_i, and therefore by Problem 8, $f(T)^*$ is a polynomial in T. By Problem 5, $f(T)$ is normal. The second statement follows from the representation

$$f(T) = f(\lambda_1)T_1 + \cdots + f(\lambda_k)T_k$$

(see Problem 4).

10. To find the eigenvalues, we must solve

$$\begin{bmatrix} \cos \psi - \lambda & -\sin \psi \\ \sin \psi & \cos \psi - \lambda \end{bmatrix} = 0,$$

i.e., $\lambda^2 - 2\lambda \cos \psi + 1 = 0$. The eigenvalues are $\cos \psi \pm i \sin \psi$. When $\lambda = \cos \psi + i \sin \psi$, the equations $Ax = \lambda x$, with $x = (u, v)^t$, reduce to $(-i \sin \psi)u - (sin\psi)v = 0$, or $u = iv$. Thus $(i, 1)$ is an eigenvector. When $\lambda = \cos \psi - i \sin \psi$, we get $(i \sin \psi)u - (\sin \psi)v = 0$, so that $v = iu$. Thus $(1, i)$ is an eigenvector. An orthonormal basis of eigenvectors is given by $(i/\sqrt{2}, 1/\sqrt{2})$ and $(1/\sqrt{2}, i/\sqrt{2})$.

Section 6.5

1. Near the end of the proof we said ... let λ be any eigenvalue of A. We need the complex numbers to guarantee that A has at least one eigenvalue (see Example 6.4.2). If A is n by n, the eigenvalues are the roots of $\det(A - \lambda I)$, which is a polynomial of degree n in λ. The key point is that every polynomial of degree at least 1 with coefficients in the field of complex numbers has at least one root. A field in which this property holds is said to be *algebraically closed*. It can be shown that the Jordan canonical form exists over any algebraically closed field.

2. (a) $S(Tx) = STx = TSx = T(\lambda x) = \lambda(Tx)$.

(b) If $x \in W$, then $Sx = \lambda x$ for some λ, so by (a), $S(Tx) = \lambda(Tx)$, hence $Tx \in W$.

(c) If $m_T(x)$ is the minimal polynomial of T, then $m_T(T) = 0$, in particular, $m_T(T)$ is 0 on W. Thus the minimal polynomial $q(x)$ of T_W divides $m_T(x)$ by (6.2.2). But by (6.2.5), $m_T(x)$ is a product of distinct linear factors, hence so is $q(x)$. Again by (6.2.5), T_W is diagonalizable. If T is unitarily diagonalizable and therefore normal, then

$TT^* = T^*T$; in particular, this holds on W, so that T_W is also normal and therefore unitarily diagonalizable.

(d) Since S is diagonalizable, there is a basis of eigenvectors of S. By (c), T is diagonalizable on each eigenspace W of S, so we may choose a basis for W whose members are eigenvectors of both T and S. If we do this for each eigenspace of S, we have simultaneously diagonalized the operators.

(e) Proceed exactly as in (d), with "diagonalizable" replaced by "unitarily diagonalizable" and "basis" by "orthonormal basis".

(f) There is a basis whose members are eigenvectors of both T and S. With respect to this basis, both T and S are represented by diagonal matrices, which always commute. Therefore $TS = ST$.

List of Symbols

Symbol	Meaning	First Appearance
iff	if and only if	4
$\vee$	or	4
$\wedge$	and	4
$\neg$	not	4
$\Rightarrow$	implies	4
$\Leftrightarrow$	equivalence	4
$\exists$	there exists	6
$\forall$	for all	6
$\in$	set membership	11
$\cup$	union	11
$\cap$	intersection	11
c	complement	11
$\subseteq$	subset	13
$\subset$	proper subset	13
$\varnothing$	empty set	13
$\setminus$	difference between sets	13
$\circ$	composition	14
f^{-1}	preimage under f	16
(a, b)	ordered pair	18
$A \times B$	cartesian product	19
$\equiv$	congruence	19, 52
$\binom{n}{k}$	combinations of k objects out of n	26
$\mathbb{Z}$	integers	40
$\mathbb{Z}^+$	positive integers	40
$\mathbb{Q}$	rationals	40
$\mathbb{R}$	reals	40
gcd	greatest common divisor	45
lcm	least common multiple	51
$\mathbb{Z}_m$	integers modulo m	54
μ	Möbius function	64
$\mathbb{N}$	natural numbers	69
$\leq_s$	equal or smaller size	74
$=_s$	same size	74
$|A|$	cardinal number of A	77
$\aleph_0$	cardinal number of a countably infinite set	77

2^A	power set of A	77
$\alpha + \beta$	cardinal addition	78
$\alpha\beta$	cardinal multiplication	78
α^β	cardinal exponentiation	80
$M_{mn}(F)$	$m \times n$ matrices over F	82
F^n	n-dimensional vectors with coefficients in F	92
$\mathbb{R}^3$	Euclidean 3-space	92
$N(A)$	null space of A	100
$R(A)$	range of A	100
$< x, y >$	inner product of x and y	108
$\|x\|$	norm of x	108
$\mathbb{C}$	complex numbers	108
$x \perp y$	x and y are orthogonal	109
$\oplus$	direct sum	117
$m_A(x)$	minimal polynomial of A	127
$c_A(x)$	characteristic polynomial of A	129
T^*	adjoint of T	132
$\mathbb{R}^2$	Euclidean plane	135

Index

abelian group, 52
additivity, 102
adjoint matrix, 89
adjoint of a linear operator, 132
algebraic structure, 52
antisymmetric relation, 21, 69
associativity, 51–52
axiom of choice, 73

basis of a vector space, 74
basis step in proofs by induction, 10, 72
Bessel's inequality, 114
bijection, 15
binary arithmetic, 41
binary operation, 51
binomial expansion modulo p, 62
binomial theorem, 32

Cantor's diagonal process, 40
cardinal arithmetic, 78–80
cardinal number (cardinality), 77
Cartesian product, 19
Cauchy-Schwarz inequality, 109
Cayley-Hamilton theorem, 130
chain, 21, 71
change of basis, 105–106
characteristic polynomial, 129
Chinese remainder theorem, 60
closure, 51–52
codomain, 14
cofactor, 88
column space, 97
column vector, 82
combination, 26
commutative ring, 52
commutativity, 51–52
complement, 11
complex vector space, 108
composite, 48
composition, 14
congruence modulo m, 19, 52
conjugate transpose, 117, 132
connectives, 1
continuum hypothesis, 80
contradiction, 5, 9
contrapositive, 8
converse, 3
coordinates, 104, 105
coset, 101
countable set, 40
countably infinite set, 40
counting infinite sets, 40–43

counting surjective functions, 39
Cramer's rule, 89

DeMorgan laws, 4, 12
derangement, 38–39
determinant, 86ff..
determinant of a linear operator, 115
diagonal matrix, 115
diagonalizable, 115, 129
difference between sets, 13
dimension of a vector space, 94
dimension theorem 100, 105
diophantine equations (linear), 57ff.
direct proof, 8
direct sum, 116–117
disjoint sets, 13
distributive law for sets, 13
distributive laws, 52
domain, 14
duality, 6

echelon form, 84
eigenspace, 116
eigenvalue, 115
eigenvector, 115
elementary row and column matrices, 83
elementary row and column operations, 83
empty set, 13
equivalence relation, 19
Euclidean algorithm, 45
Euler phi function, 37, 61, 63, 66–67
Euler's theorem, 62
existence of the Jordan canonical form, 141
existential quantifier, 6

Fermat's (little) theorem, 62
field, 52
finite Markov chain, 145
functions, 14ff.

generalized eigenvectors, 126
generalized permutations, 28
Gram-Schmidt process, 111
greatest common divisor, 45
greatest lower bound, 23
group, 52

Hermitian matrix, 118, 133
homogeneity, 102
homogeneous linear equations, 99, 100

idempotent operator, 133
identity (additive, multiplicative), 51–52
identity function, 75
if and only if, 3–4

image, 17, 102
implies, 2–3
inclusion relation, 22
index set, 71
induction hypothesis, 10
injective, 15
inner product, 108
integral domain, 52
intersection, 11–12
invariant subspace, 117, 143
inverse (additive, multiplicative), 51–52
inverse of a matrix, 87
invertible list (conditions equivalent to invertibility of
    a matrix), 98–99, 102–103
isomorphism, 101

Jordan canonical form, 123ff.
Jordan block, 123

kernel, 102

Lagrange interpolation, 140
Laplace expansion, 88
least common multiple, 51
least squares, 114
least upper bound, 22
length, 108
lexicographic (dictionary) ordering, 72
linear operator, 106, 114ff.
linear transformation, 102
linearity, 102
linearly depenedent, 74, 92
linearly independent, 74, 92
lower bound, 23

mathematical induction, 9–11
matrix, 81ff.
matrix that represents a linear transformation, 104
maximal chain, 71
maximal element, 21, 72–73
maximum principle, 71
minimal polynomial, 127ff.
minor, 89
Möbius function, 64
Möbius inversion formula, 65
multinomial theorem, 33
multiple count, 31
multiplication rule, 25
multiplicative function, 67

necessary condition, 2–3
nonhomogeneous linear equations, 100
nonnegative definite, 119
norm, 108
normal equations, 114
normal matrix, 135
normal operator, 134
null space, 100

one-to-one, 15
only if, 2

onto, 15
ordered pair, 18
ordered samples with replacement, 26
ordered samples without replacement, 26
ordered n-tuple, 19
orthogonal (perpendicular), 109
orthogonal complements, 134, 136
orthogonal diagonalization, 119
orthogonal direct sum, 138
orthogonal matrix, 118
orthonormal basis, 110
orthonormal basis of eigenvectors, 117

parallelogram law, 110
partial ordering, 21, 69
particular solution, 101
partition, 20, 31, 39–40
Pascal's triangle, 34
perfect number, 67–68
permutation, 26
PIE, 34ff.
polarization identity, 113
polynomial in a linear operator, 114
positive definite, 119
power set, 77
preimage, 16
prime, 48
Principle of inclusion and exclusion, 34ff.
projection, 110
projection operator, 112, 133, 138–139
projection theorem, 112
proof by cases, 9
proofs, 8ff
proper subset, 13.
propositions, 1
Pythagorean theorem, 110

quadratic form, 119
quantifiers, 6

range, 100
rank, 98, 100, 103
rationals are countable, 40
real vector space, 108
reals are uncountable, 40–41
reflexive relation, 20, 21, 69
relations, 18ff.
relatively prime, 37, 56
relatively prime in pairs, 60
residue, 52
residue class, 52
restriction, 75
ring, 52
rotation, 135, 140
row space, 97
row vector, 82

Schröder-Bernstein theorem, 75
self-adjoint operator, 133
set theory, 69ff.

sets, 11ff.
sieve of Eratosthenes, 50
similar matrices, 106, 115
simultaneous diagonalization, 143
skew-symmetric matrix, 85
spanning (generating) set, 93
spectral theorem for normal operators, 139
spectral theorem for self-adjoint operators on a real
   space, 139
standard basis, 93
stars and bars, 27
Steinitz exchange, 93
stochastic matrix, 145
strong induction, 69–70
strong induction hypothesis, 70
stronger hypothesis, 22
subset, 13
subspace, 97
subspace spanned by a set of vectors, 97
sufficient condition, 2–3
superposition principle, 102
surjective, 15
Sylvester's law of inertia, 121
symmetric relation, 21
symmetric matrix, 85

tautology, 5, 9
total ordering, 21, 69
transfinite induction, 71, 72

transition matrix, 145
transition probabilities, 145
transitive relation, 20, 21, 69
transpose of a matrix, 85
triangle inequality, 109
truth tables, 1

uncountable set, 40
uncountably infinite set, 40
union, 11–12
unique factorization theorem, 49
unit vector, 109
unitary diagonalization, 119
unitary matrix, 118
universal quantifier, 6
unordered samples with replacement, 27
unordered samples without replacement, 26–27
upper bound, 22, 73

vacuously true, 5
vector space, 73, 92
Venn diagrams, 12

weaker hypothesis, 22
well-ordering, 69
well-ordering principle, 70
without loss of generality, 94

zero-dimensional space, 95
zero-divisor, 52
Zorn's lemma, 73